for 9th and 10th grade levels

Geometry Challengers

for high achievers

A Resource Book of over 200 problems from the elements of geometry
to lines, polygons, triangles, analytic geometry, polyhedrons, solids, trans-
formations, and introduction to vectors, basic mechanics,
and complex numbers.
with easy-to-follow solutions and answers as learning guides
toward skill-building and enrichment

**for teachers, high achievers
and all forward-looking students**

**also for challenge and fun
at party time
and refresher sessions
for parents and grown-ups**

CESAR G. QUEYQUEP, Ph.D.

*Department of Engineering
Purdue University Calumet
Hammond, Indiana, USA*

*Department of Mathematics
Bishop Noll Institute
Hammond, Indiana, USA*

TRAFFORD
PUBLISHING™

Canada * USA * UK * Ireland

Note for Librarians: A cataloguing record for this book is available from Library and Archives
Canada at www.collectionscanada.ca/amicus/index-e.html
ISBN 1-4120-8914-X

PUBLISHING™
Offices in Canada, USA, Ireland and UK

Book sales for North America and international:
Trafford Publishing, 6E–2333 Government St.,
Victoria, BC V8T 4P4 CANADA
phone 250 383 6864 (toll-free 1 888 232 4444)
fax 250 383 6804; email to orders@trafford.com
Book sales in Europe:
Trafford Publishing (UK) Limited, 9 Park End Street, 2nd Floor
Oxford, UK OX1 1HH UNITED KINGDOM
phone 44 (0)1865 722 113 (local rate 0845 230 9601)
facsimile 44 (0)1865 722 868; info.uk@trafford.com
Order online at:
trafford.com/06-0670

10 9 8 7 6 5 4 3 2

Dedicated to . . .

Caroline 1 and Lili

and Rose, Maribel, and Ces
      who constantly set their sights
above life's distant horizons

. . . high up toward the unreachable realms
      of rainbows and stars.

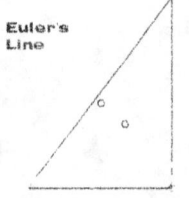

Euler's
Line

Foreword

The better caption for these few paragraphs should be *Why I Wrote This Book.*

In my many years of teaching mathematics in the American classroom, I observed a striking feature of learners, young and old, that seemed true for not only the sharper upper level but also the mainstream group.

It is safe to say that the upper third, or usually, the upper brighter half of the class understood concepts and lessons well if these were explained well, clearly, and unconfusingly.

Beyond this lay a visible hunger for something more than the usual homework and classwork. The lessons and homework assignments were discussed well in class, in both average and high-level ranks of students. There was a yearning for extra work, definitely a need to stretch the opportunity to apply learned concepts. They like the idea of challenge.

Especially if extra credit work was aptly rewarded. Challenge problems were well-received and satisfying. They brought problems, challenge problems, home and came back with answers, the right answers. Clearly, challenge problems are not only skillbuilders, they reinforce learning and enhance the power of recall as well.

In all, challenge problems are such resources that broaden the capacity to think and learn.

The problems in this book have been conceived in the hope that they will provide an answer for the inquiring mind, the adventurer, and the thinker in most of us.

I had been told by many of my students that the sparks that lit the students' minds when absorbed with the challenge problems they brought home from school, came from discussions with family members and parents eager to face the challenge themselves. The problems were even brought around water coolers and drinking fountains in steel mills and factories where students' parents worked, and the discussions there were lively and intelligent.

Challenge problems are like riddles and puzzles that are fun to sit down for. They light up discussions and bring fun at party time. Grown-ups love to relive school memories especially the struggles and triumphs with brain teasers in the math classroom of long ago. Everyone agrees that challenge problems are indeed what they are about ... challenge, a challenge that sharpens the mind and pushes forward the frontiers of knowledge.

Cesar G. Queyquep, Ph.D.
Ryland Oasis
Menifee, Southern California

About the author

Dr. Cesar G. Queyquep earned his aeronautical engineering degree (top 10) from Feati University, his MS degree in mathematics education from Purdue University (as an NSF scholar), and his doctorate degree in education, summa cum laude, from Madison University. He was valedictorian of his graduating class at the Urdaneta Provincial East High School, now the Urdaneta City National High School.

Dr. Queyquep served on the faculty of the Department of Engineering of Purdue University Calumet (Northwest Indiana campus) as professor of Computer-Aided Design, Engineering Drawing, and Engineering Design. While connected with Purdue, he was also on the faculty of the Department of Mathematics at nearby Bishop Noll Institute.

At Bishop Noll Institute, he taught honors geometry, physics, algebra 1 and 2, industrial design, and mechanical drawing. He was Faculty Sponsor for the Math Club for three decades, Director of the Annual Mathematics Tournament for the Diocese of Gary and neighboring Chicagoland schools. He was also the Faculty Sponsor and Coach of the Bishop Noll JETS Academic Team which won three successive First Place Trophies in the Annual JETS Academic Tournaments at Valparaiso University and Purdue University Calumet. He was inducted into the Bishop Noll Institute Hall of Honor for outstanding service in the field of education to the Bishop Noll Community.

He authored two books published by the Kendall-Hunt Publishing Co., namely: "The Essential AutoCAD" (for engineering students in computer-aided design) and "The Essential Engineering Graphics Concepts" for engineering students. Both books were used at Purdue University Calumet, Ivy Tech State College Northwest, Bishop Noll Institute and other area schools, He was also a contributing author (he wrote the Research Project Report on "The Mathematics of Flight") in the Student Merit Awards book published by the National Council of Teachers of Mathematics in Reston, Virginia.

He taught college physics as a graduate teaching assistant at Western Michigan University. At Feati University, his alma mater, he was an assistant professor teaching aerodynamics, engineering physics, and theory of flight to aircraft maintenance, engineering and flying students, and later became Head of the Department of Aeronautical Engineering.

Table of Contents

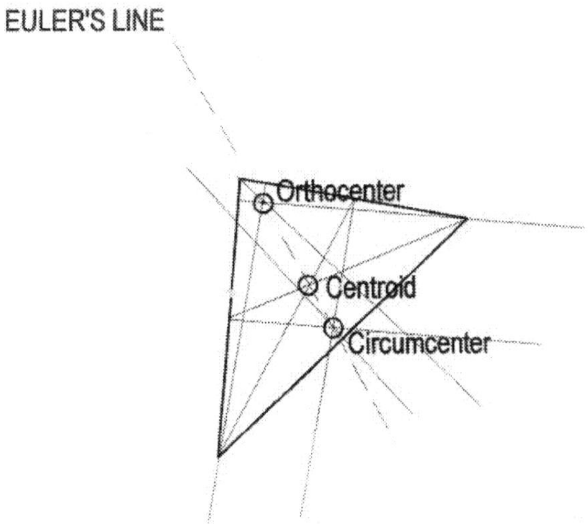
EULER'S LINE
Orthocenter
Centroid
Circumcenter

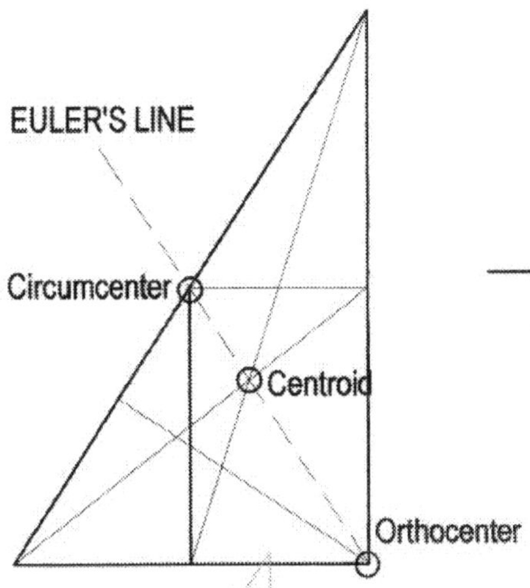
EULER'S LINE
Circumcenter
Centroid
Orthocenter

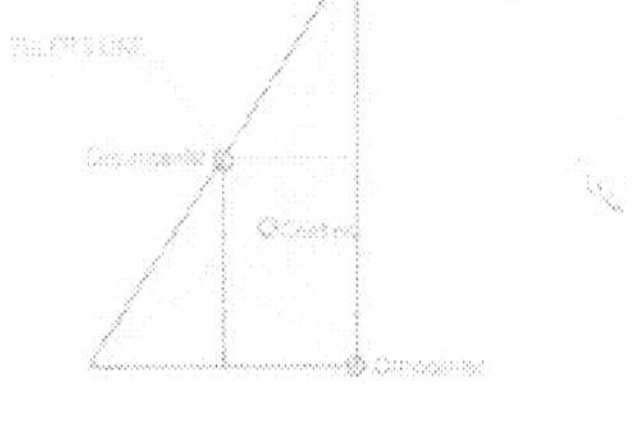

Part One

Problems

Points, lines, distances

Runners S and T are 2000 feet apart. They run toward each other at the same speed, and after meeting at the 1000 feet mark, each one turns around and heads back to their respective starting points. Each one goes back and forth on 10 back and forth trips, but on each return trip each runner covers only half the distance traveled in going.

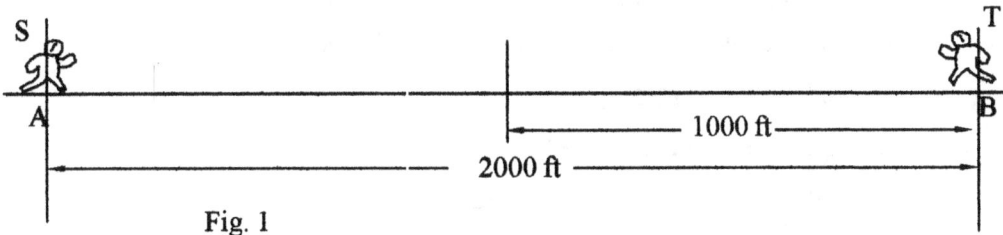

Fig. 1

1. After 5 back-and-forth trips, how far apart are S and T?

2. How far did each runner travel after making 5 back-and-forth trips?

2A. Will S and T ever meet?

3. Points A and B have coordinates -8 and -52, respectively, on the number line. Point C is halfway between A and B, point D is at 1/3 AB from B, E is at 1/3DB to the right of D, and F is at 1/5 DE to the right of E. Locate (give the coordinates of) A and F, and determine the distance between them.

4. A painter has to draw a spiral line from the base to the top of a 20 ft cylindrical pole with a diameter of 3 ft. The line drawn around the cylinder makes an angle of 45 degrees with the ground. How many complete turns will the line make until it reaches the top of the pole?

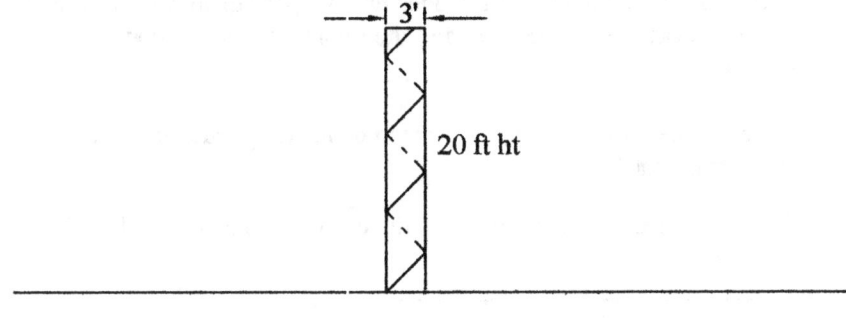

Fig. 2

5. If a line does not intersect a plane, could the line and the plane intersect a second line at some point outside the plane? Explain why or why not.

6. Is it possible for ten planes A, B, C, D, E, F, G, H, I and J to intersect in 15 segments? Draw a figure, label all the planes and the intersections, and explain.

Angles and angle measurements

7. A Ferris Wheel has 12 radial beams around a circle. The beams are numbered 1 to 12. If the wheel makes one turn every half-minute, how much time will it take beams 5 and 6 to pass by a fixed point A on the wheel support?

8. If the wheel in prob. 7 has 14 beams instead of 12, express in radians the difference in size between the angle formed by beams 5 and 12 and the angle formed by beams 1 and 6.

A bicycle wheel has 32 spokes, all equally spaced around a circle.

9. Find the measure of the angle between spokes 11 and 27.

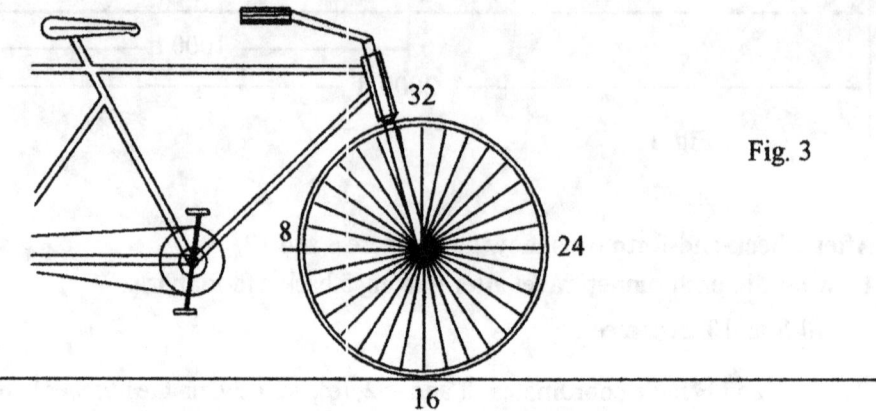

Fig. 3

10. If 10 spokes pass a given point in 0.5 second, how fast is the wheel turning in turns per minute?

11. If the spokes were painted red, white, blue and green consecutively, how fast should the wheel turn so that three consecutive white spokes would go past a fixed point A on the wheel supports in 0.9 second?

The counter-rotating blades of a big two-blade lawn mower both have starting positions on the positive y-axis. Both blades turn in opposite directions, with the left blade turning clockwise at 540 rpm, and the right blade turning counterclockwise at 540 rpm as well.

12. What is the measure of the angle formed by the two blades at the end of 1/16 second from start time?

13. In how many seconds after start will the blades form an angle of 90 degrees?

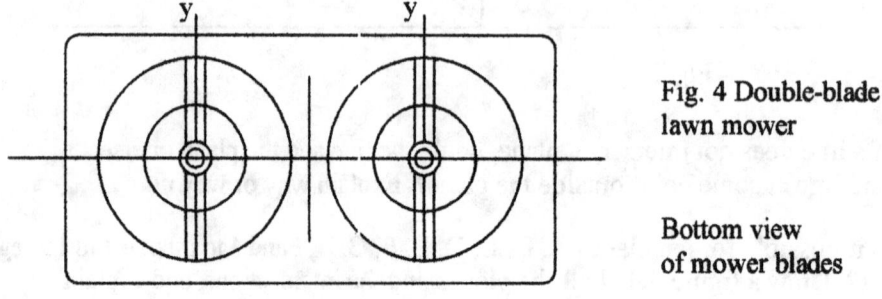

Fig. 4 Double-blade
lawn mower

Bottom view
of mower blades

Fig. 5

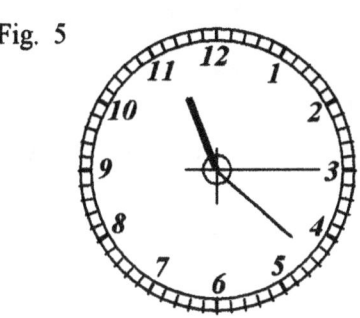

14. The clock in Fig. 5 above shows the time as 11:20:15. What is the measure of the angle between the hr and sec hand, and the angle between the min and sec hand?

15. A clock shows the time as 8:20:30. Determine the angle (a) between the hr and sec hand, and (b) between the min and sec hand.

16. At what time between 10 pm and 11 pm will the minute and hour hands of a clock form an angle of 90 degrees?

17. Through how many radians does the earth turn about its axis from 12 noon to 3:12 pm?

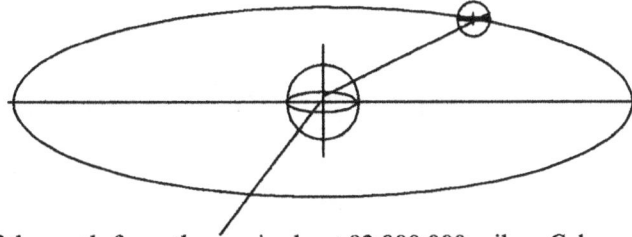

Fig. 6

18. The mean distance of the earth from the sun is about 92,000,000 miles. Calculate the angle, in radians, traveled by the earth around the sun from 12 noon to 3:12 pm.

19. In Fig.7:　　　PR bisects ∠ SPQ
　　　　　　　　　PS bisects ∠ TPQ
　　　　　　　　　PT bisects ∠ UPQ
　　　　　　　　　PU bisects ∠ VPQ
　　　　　　　　　PV bisects ∠ WPQ, and
　　　　　　　　　PW bisects ∠ XPV

Find the measure of ∠ APB if ∠ BPQ = ∠ TPQ

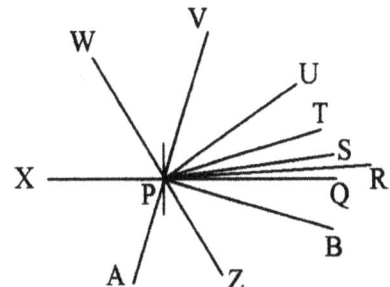

Parallel lines, polygons

20. Given: MB and AB are bisectors of ∠AMN and ∠RAM, respectively.
(Fig. 8)   ∠CAD = 90 - a,   ∠VMW = a

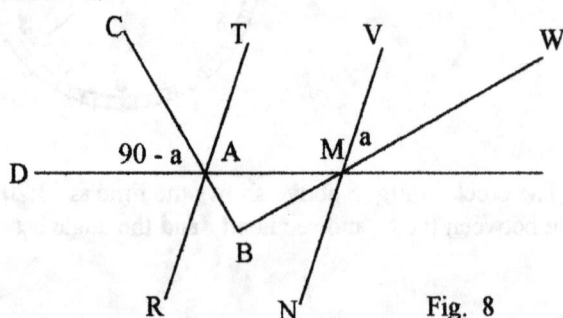

Fig. 8

    Prove: RT ‖ VN

21. Is it possible for an exterior angle of a regular polygon to have a measure of 21 degrees? Explain.

Find the measure, in terms of n(the number of sides of a regular polygon), of the angle formed by the bisectors

22. of the adjacent interior angles of a regular octagon; and

23. of the adjacent interior angles of a regular decagon.

24. Given, equilateral △STU. TV bisects ∠STU and M is the midpoint of ST.

Fig. 9

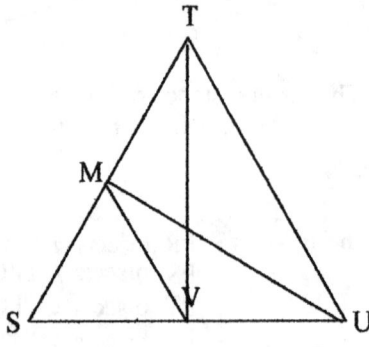

Find the measure of ∠UMV.

25. Given, obtuse △ACB (Fig. 10)
∠A = 2∠B
∠C = 3∠A

CM bisects ∠C, and TM bisects ∠CMB.
Find the measure of ∠1.

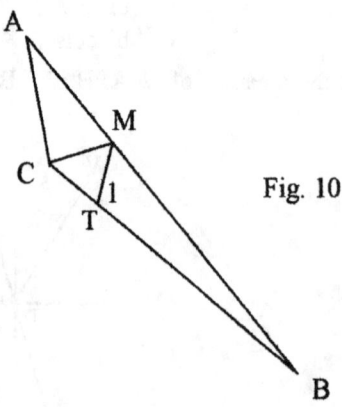

Fig. 10

26. (Fig. 11). AB and CD lie on the sides of ∠ AVC, where V is the inaccessible vertex. Without extending the sides AB and CD of the angle, explain how the angle could be bisected.

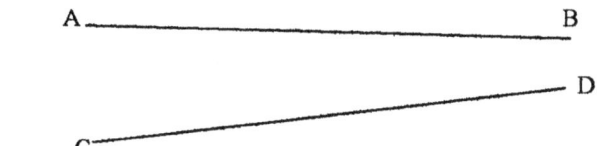

Fig. 11

Perpendicularity, right angles

27. Fig. 12 shows a pyramid V-KLMN.

Fig. 12

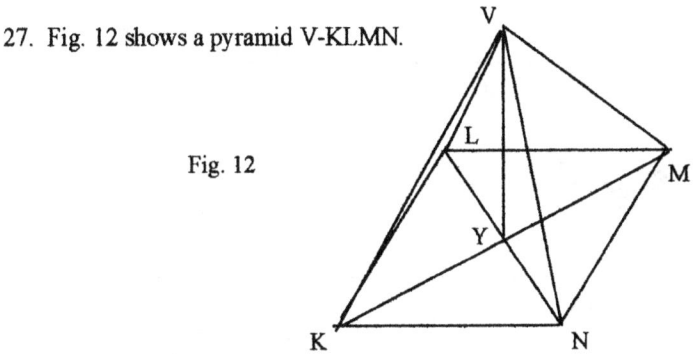

The base is a square and each side (lateral face) is an equilateral triangle. KM and LN are the diagonals of the base, and the altitude VY passes through the center of the base.

Prove: NY ⊥ KM.

28. Fig 13 shows a regular pyramid V-KLMN. The base is a square and each of the lateral faces is an equilateral triangle.

Fig. 13

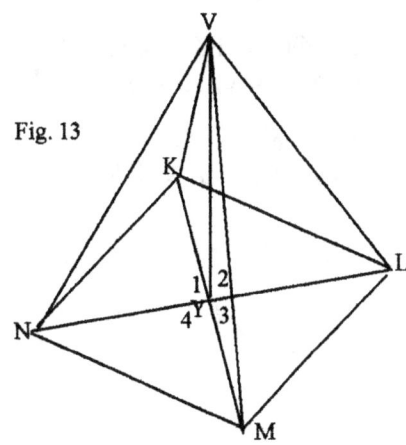

Prove: VY ⊥ KM.

29. Determine the area of △PQR whose perimeter is 90 inches, if PQ = 3x - 10, QR = 2x + 10, and PQ = 6x - 20. (Fig. 14)

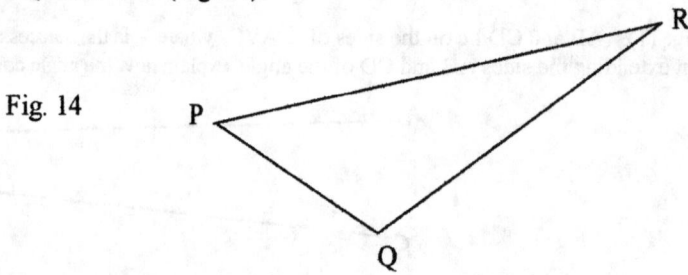

Fig. 14

30. The perimeter of △STU is 126 inches. SU is twice as long as ST, and TU is 1.5 times as long as ST. Find the area of the triangle. (Fig. 15)

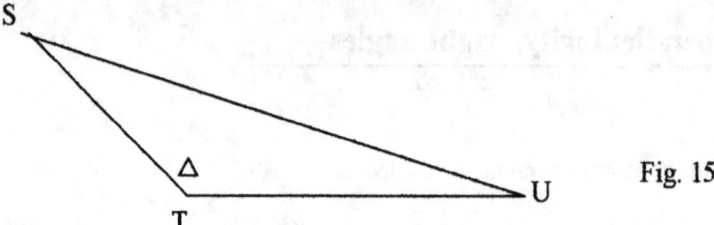

Fig. 15

Congruent triangles

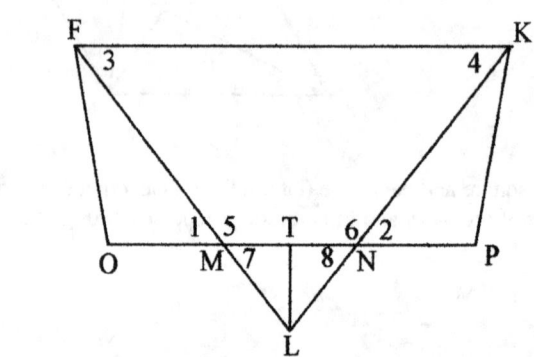

Fig. 16

31. Given, △ FKL and △ MNL are isosceles triangles. (Fig. 16) OM =PN, ∠3 = ∠4, ∠3 and ∠4 are complementary. Prove: △FOM ≅ △KPN.

32. Given: FK ‖ OP, ∠3 = ∠4. (Fig. 16). △MLN is isosceles, and LT is a median. Prove: △ LTN ≅ △LTM.

33. (Fig. 17)   Given: PM ⊥ US, RN ⊥ TS, PT = UR, RN = PM, SN = SM. Prove: △ PTS ≅ △RUS.

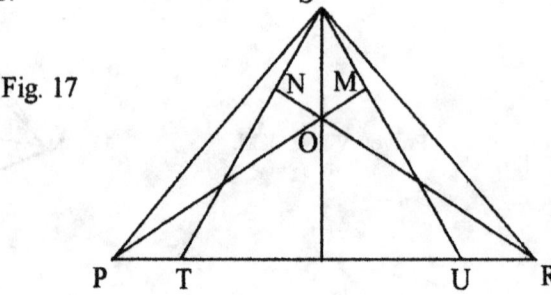

Fig. 17

34. In Fig. 18, EC ⊥ BC, AB ⊥ CA, AB ‖ DE, and AB = CD.
Prove: △BAC ≅ △DCE.

Fig. 18

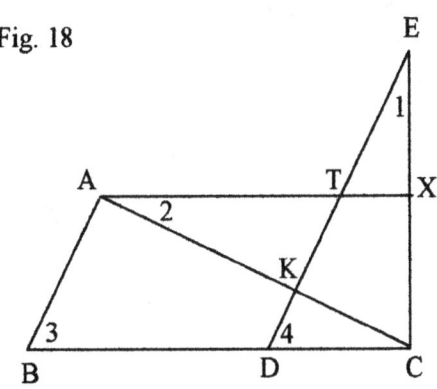

35. Referring to Fig. 18, AX ‖ BC, EC ⊥ BC, AC ⊥ AB, ED ‖ AB, and
EC = AC.   Prove: AB = CD.

36. (Fig. 18) Given: EC ⊥ BC, AC ⊥ AB, ED ‖ AB, ED = BC.
Prove: △BAC ≅ △DCE.

Quadrilaterals

37. In Fig. 19,   △AMC ≅ △BRN,   ∠3 ≅ ∠2, PS ≅ QR, PS ‖ QR.
Prove: ∠1 ≅ ∠B + ∠SQR.

Fig. 19

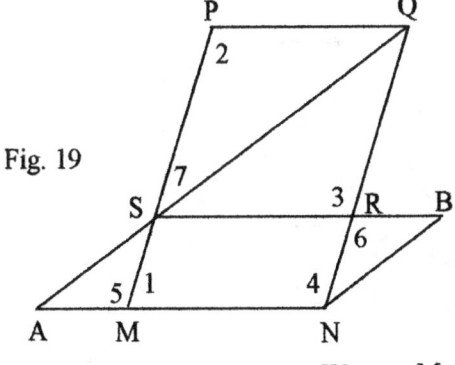

38. Given, the cube in Fig. 20

Fig. 20

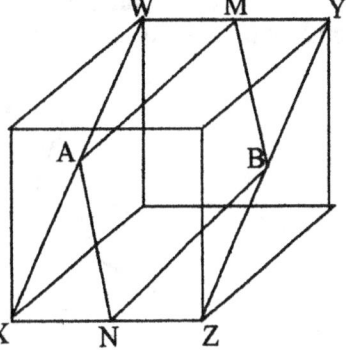

A and B are the midpoints of WX and YZ respectively. M and N are the midpoints of WY and XZ respectively. Express the perimeter of MBNA in terms of the length of an edge of the cube. (Fig. 20)

39. Prove that the length of a median MN of any triangle (except the isosceles and the equilateral) is greater than the length of its altitude ME. See Fig. 21. Both should be altitude and median to the same side of the triangle.

Fig. 21

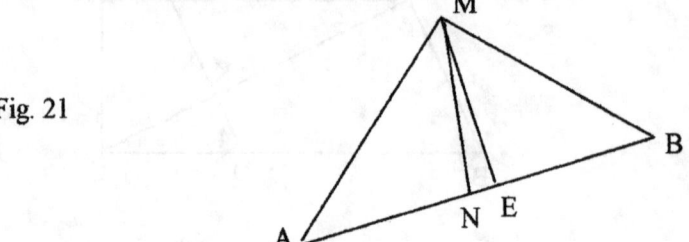

40. Given: ▱ DEFG.   EF and GD bisect ∠E and ∠G respectively. (Fig. 22)
Prove: CENG is a ▱.

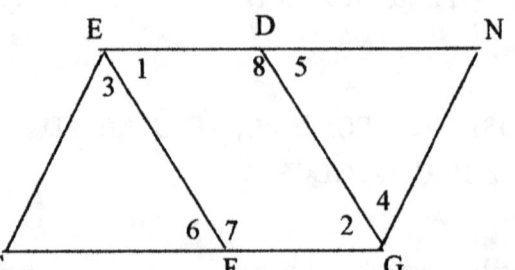

Fig. 22

41. Given, the cube in Fig. 23. A and B are the midpoints of MN and KL respectively.
Prove: RETA is a rhombus.

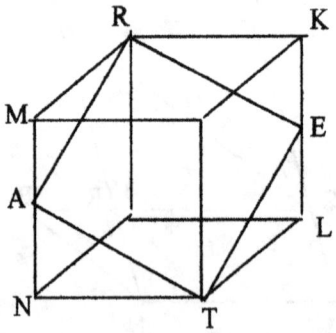

Fig. 23

42. V, M, T, and U are the midpoints of the sides of the upper base (Fig. 24), and R, O, B, and Y are the midpoints of the sides of the lower base. K is the midpoint of VR, X is the midpoint of TB.

Fig. 24

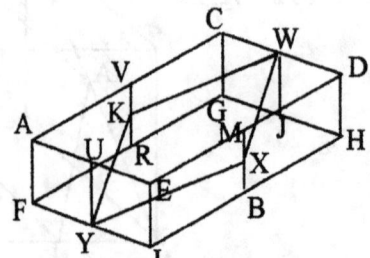

Prove: KWXY is a rhombus.

Ratio and similarity

43. A uniform beam is loaded at each end with weights of 150 lbs and 250 lbs. Find the ratio of the distances d_1 and d_2 from the fulcrum (point of support) so that the beam would be in equilibrium (or that it would balance horizontally).

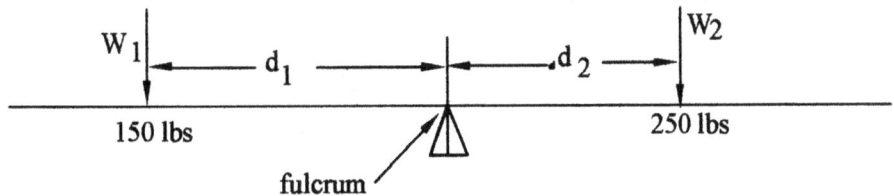

NOTE: The right end load tends to turn the beam counterclockwise about the fulcrum F clockwise. The turning effect of the load on the beam is termed "moment of force" and is equal to the product of the force F(or W in this case since weight is also a force) and the distance d of the line of action of the force from the fulcrum. A counterclockwise moment is taken as positive, and the clockwise moment negative. For the beam to balance horizontally, the moments on the beam must have an algebraic sum of zero.

44. The freezing point of water in the Fahrenheit scale is 32 degrees and the boiling point is 212 degrees. The freezing point of water in the Centigrade scale is 0 degree, and the boiling point is 100 degrees.
From these given relationships, derive the formulas below:

$$C = 5/9\ (F\ -\ 32)$$
$$F = 9/5C\ +\ 32$$

Fig. 26

45. Fig. 27 shows trapezoid XYZS with median VW.

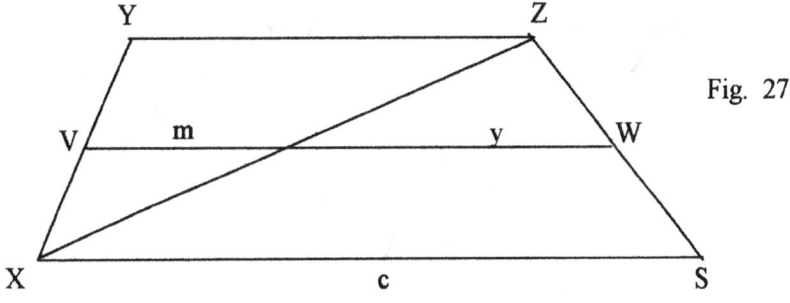

Fig. 27

Derive the formula for determining the length of the median m (see Fig. 27) of a trape-zoid.

$$m = 1/2\,(a + b)$$

where a and b are the upper and lower bases of the trapezoid.

46. Given, the isosceles trapezoid KLMN below:

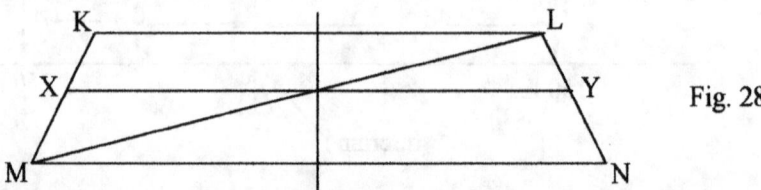

Fig. 28

If the length of the median XY is 16x and that of the base MN is 24x, express the length of the upper base KL in terms of the length of the diagonal NL. The altitude of the trape-zoid is 10 inches.

47. Given, the cube in Fig. 29.
Show that AB(MV) = MN(AV).

Fig. 29

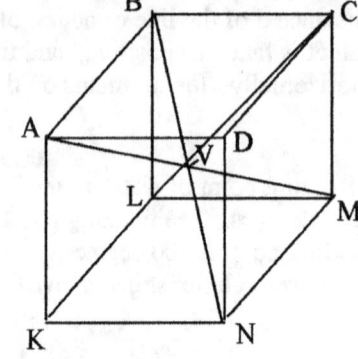

48. Given, right triangle ABC (Fig. 30).

Fig. 30

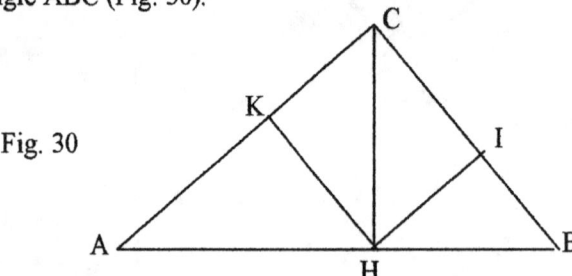

CH ⊥ BA, HI ⊥ BC, HK ⊥ CA, BC = 15, AC = 20, BA = 25. Find the lengths of CH, HI, HK, and KI.

49. Fig. 31 shows a right triangle ACB with altitude CD and median CN.
NM ⊥ BC, AC = 5, AB = 13.

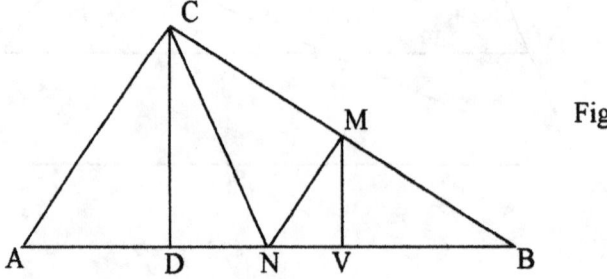

Fig. 31

The Pythagorean Theorem

50. Explain Garfield's proof of the Pythagorean Theorem. Use the figure below (Fig. 32), auxiliary lines if needed, and show the derivation of the Pythagorean formula

$$c^2 = a^2 + b^2.$$

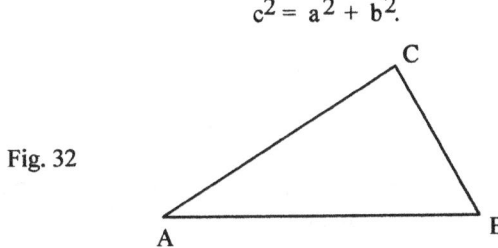

Fig. 32

51. The sides of a triangle are 10 inches, 12 inches, and 18 inches (Fig. 33). Find the length of the altitude to the longest side.

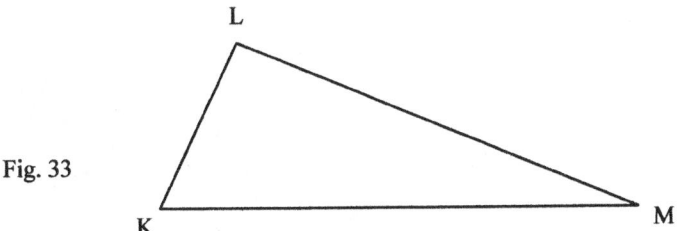

Fig. 33

52. Fig. 34 shows a cube an edge of which measures 10 inches. T, S, M, and N are the midpoints of the sides. Determine the perimeter of quadrilateral TSNM.

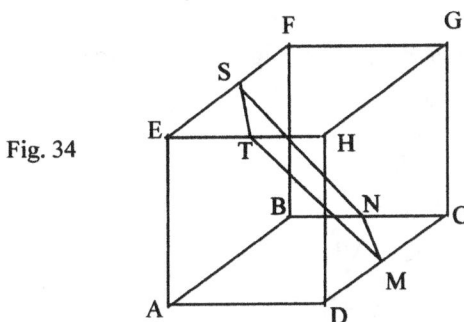

Fig. 34

53. Referring to Fig. 34 above, determine the perimeter of quadrilateral TSNM if a side of the cube is 18 inches instead of 10.

54. An angle of rhombus STUV (Fig. 35) measures 120 degrees. If the length of the longer diagonal is 12 inches, find the perimeter of a similar rhombus with a diagonal of 18 inches.

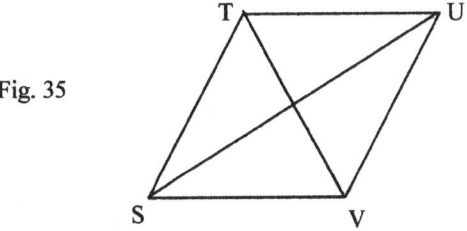

Fig. 35

55. Using the two given squares in Fig. 36 below, derive the Pythagorean formula

$$c^2 = a^2 + b^2$$

Fig. 36

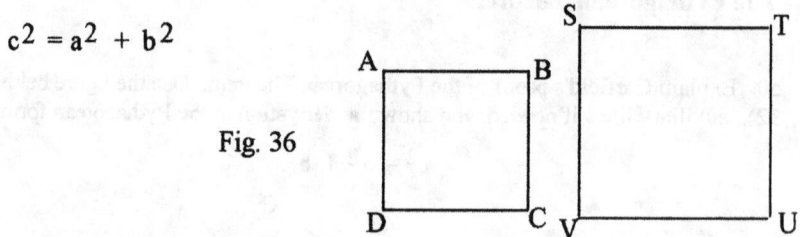

Hint: Place (inscribe) the smaller square inside the bigger.

Given, the regular pyramid in Fig. 37, the base of which is a regular dodecagon inscribed in a circle with a diameter of 10 inches. The altitude VY, 12 inches long, passes through the center Y of the base.

56. Determine the area of △BVC.

Fig. 37

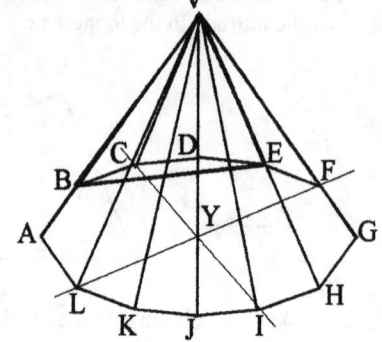

57. Determine the area of △ AEI (Fig. 37). Refer to the given in Prob. 53.

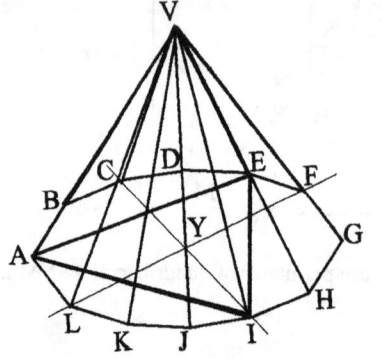

Fig. 37

58. Solve for the area of △AGI. Refer to the given in Prob. 53.

Fig. 37

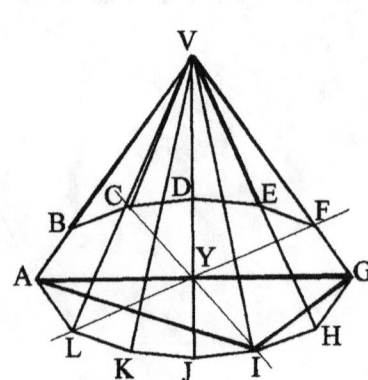

Trigonometry

59. Out in the open, an observer sees a train at point A (Fig. 38). Ten seconds later, he sees the same train at point B.

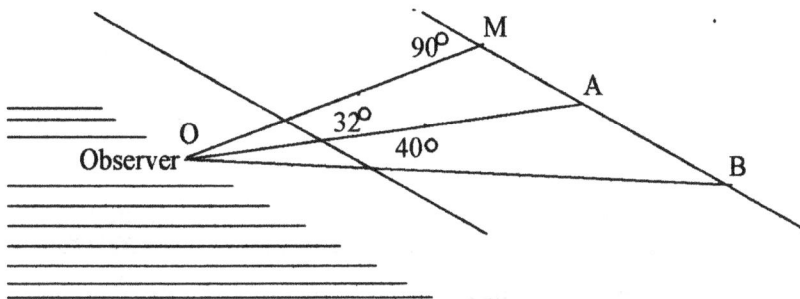

Fig. 38

Assuming that the train was traveling uniformly, how fast was it going if the distance OM from the observer to M is 1000 ft?

60. Referring to problem 59, find the speed of the train if it took 15 seconds to travel the distance from A to B.

The trajectory (path) of a missile makes an angle of 60 degrees above the horizontal. After traveling a mile on a straight trajectory (gravity effect neglected), the missile shifts to a new course 42 degrees above the horizontal., goes on half a mile along the flight path, then drops vertically downward. (Fig. 39)

61. If the missile's muzzle velocity (velocity at the instant of firing) is 15,000 ft/sec, what maximum altitude did it reach?

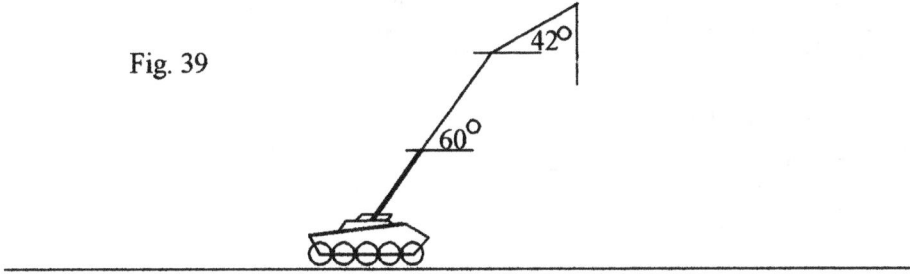

Fig. 39

62. If the angle of elevation of the missile's trajectory in prob. 61 is 55 degrees as it travels a horizontal distance of one mile, after which the trajectory angle reduces to 38 degrees above the horizontal and travels a horizontal distance of 1/2 mile before finally dropping vertically downward, how much altitude did the missile attain?

63. Referring to prob. 61, determine how far away horizontally the missile has traveled at the end of 8 seconds.

64. A gun emplacement 10 miles away horizontally shoots, at an angle of elevation of 30 degrees above the horizontal, a shell that scores a direct hit at a missile that is fired vertically upward. The shell speeds along its flight path at 600 ft/sec and lies on the same vertical plane as the missile. Find the velocity of the missile at the instant of firing. Refer to Fig. 40.

Fig. 40

65. In prob. 64, if both shell and missile were fired at precisely the same instant, at what altitude would a direct hit on the missile occur, assuming zero gravity effect and neglecting air friction?

66. In prob. 64, at what altitude would a direct hit occur at the missile, if the missile speed vertically upward is 800 ft/sec?

67. In prob. 64, at what altitude would a direct hit occur at the missile, if the missile was fired 1/2 second after the firing of the shell?

68. Using Fig. 41, derive the equation for the Law of Cosines, namely

$$c^2 \quad = \quad a^2 \quad + \quad b^2 \quad - 2ab\ cosC$$

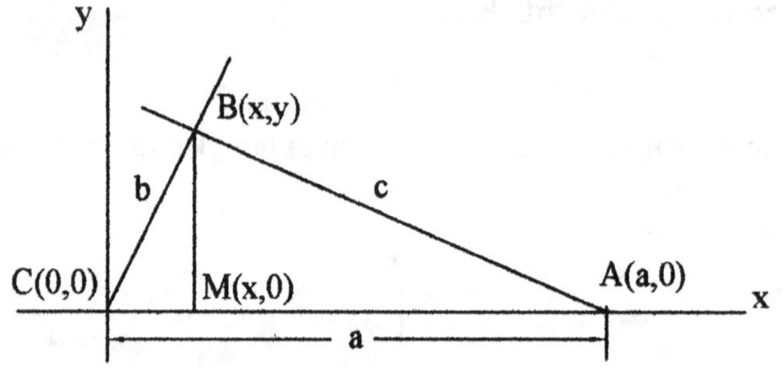

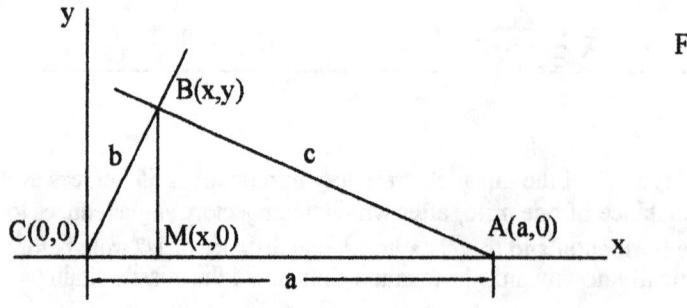

Fig. 41

69. Prove, using Fig. 42, that

$$c^2 \quad = \quad b^2 \quad - \quad a^2 \quad + \quad 2ad$$

Fig. 42

Coordinate Geometry

70. Given, the following points:

M(6,0,6)   C(6,6,6)
N(6,0,0)   B(6,6,0)
O(0,0,0)   A(0,6,0)
P(0,0,6)   D(0,6,6)

Connect the given points to form a rectangular solid. Assuming it is made of non-porous material, and if the measurements are all in inches, how many cubic feet of water would the solid displace when submerged in fresh water?

Fig. 43

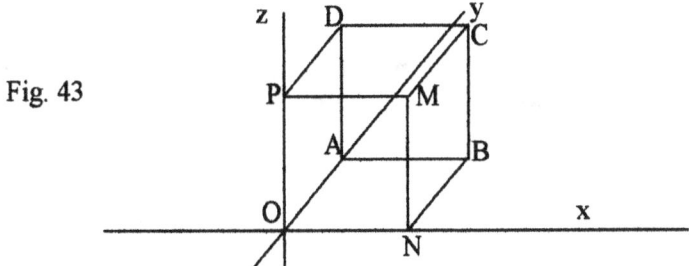

71. Given, the following points in space: O(0,0,0), P(12,0,0), Q(12,12,0), R(0,12,0) S(0,0,12), T(12,0,12), U(12,12,12), and V(0,12,12). Connect the points to form a rectangular tank, then fill halfway with fresh water. If the non-porous solid in prob. 70 is immersed into the tank, by how many inches will the water level rise in the tank? (All dimensions are in inches).

72. A(10,0), B912,2) and C(4,10) are the vertices of triangle ABC.

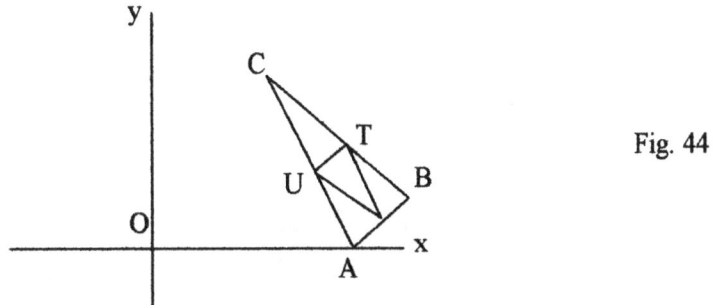

Fig. 44

S, T, and U are the midpoints of the sides of triangle ABC. Find the ratio of the perimeters and the areas of the two triangles ABC and STU.

73. The center of a circle is at C(4,6). A chord (a segment whose endpoints lie on the circle) of the circle 6 inches long is bisected at T(4, 10). Find the area of triangle ABC.

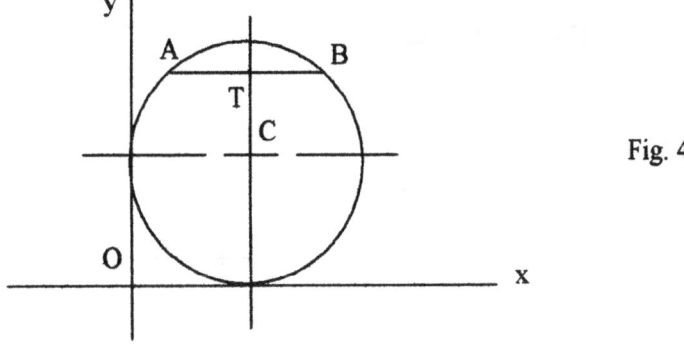

Fig. 45

74. Explain why the slope of the vertical line AB in Fig. 46 is <u>undefined.</u>

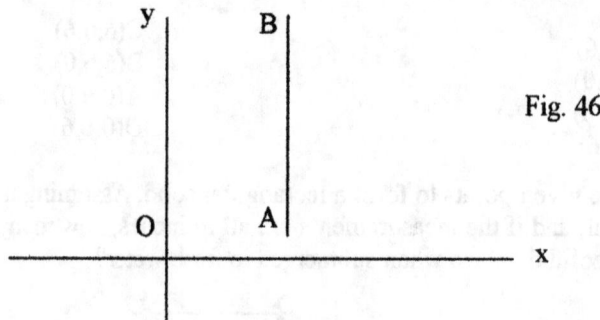

Fig. 46

75. Instead of showing that the slopes of PQ and QR are equal, use another method to prove that P(4,4), Q(8,8) and R(12,12) are collinear.

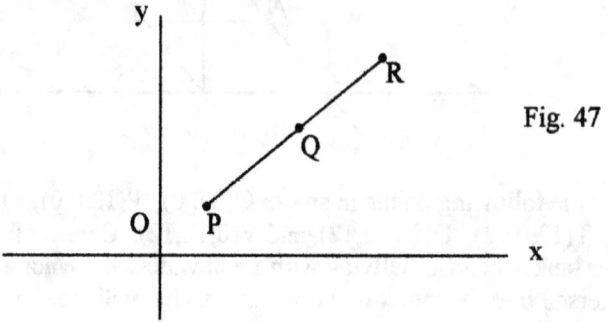

Fig. 47

76. Given, the line ST in Fig. 48. Determine the equation of the line which contains all points that are equidistant from S and T.

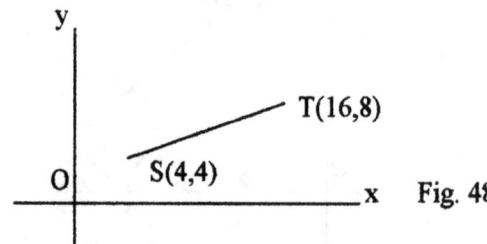

Fig. 48

77. What is the equation of the line passing through the intersection of the lines

$$4x - 6y = 10$$
$$2x + 4y = 8$$

and is parallel to the x-axis?

Fig. 49

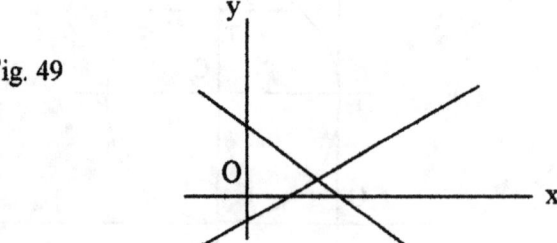

78. L(2,6), M(8,4), and N(6,10) are the vertices of △ LMN. Write the equation of the line through the intersection point of the medians of the triangle and parallel to side MN.

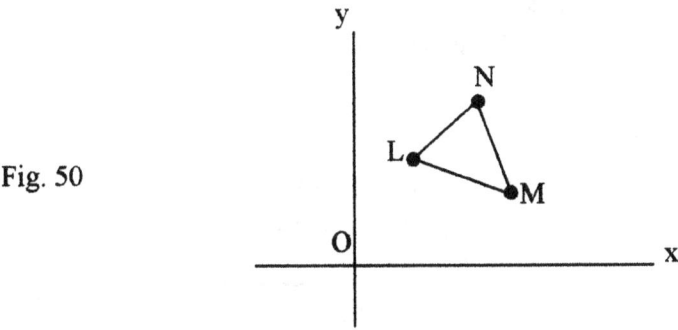

Fig. 50

79. △ABC has vertices A(6,2), B(14,6) and C(10,10); AM, CL and BK are the medians whose endpoints are connected to form △ KLM. KL intersects AM at S, LM intersects BK at T, and MK intersects CL at U. Find the coordinates of S, T, and U.

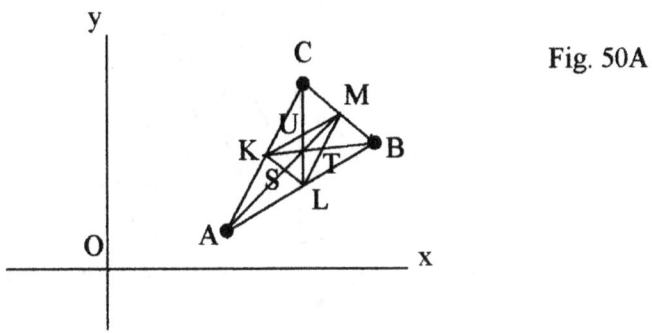

Fig. 50A

80. AE, EF, and CD are the altitudes of △ ABC. The endpoints of the altitudes are connected to form △DEF.

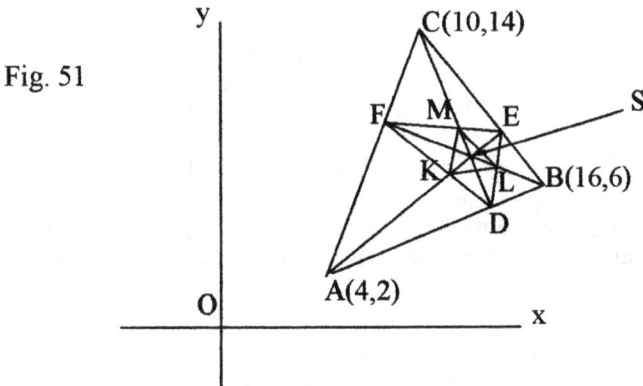

Fig. 51

Solve for the coordinates of the orthocenter S, the inersection point of the altitudes.

80a. In prob.80, the sides of △ DEF intersect the altitudes at K, L, and M. Solve for the perimenter of △KLM.

80b. Referring to prob. 78, write in the y form the equation of the line through S parallel to the side AB.

Circles

81. Given, the equilateral triangle STU and the inscribed circle with center at S. Each side of the triangle is 9 inches long.

Solve for the coordinates of the intersection points F and H.

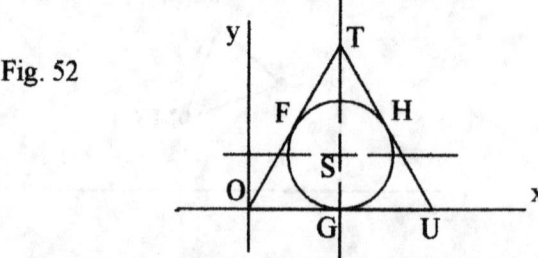

Fig. 52

82. The three congruent circles in Fig. 53 have centers at the vertices of the equilateral triangle ORQ. Each side of the triangle measures 12 inches. Determine the coordinates of the intersection points K and L in two ways and compare the results.

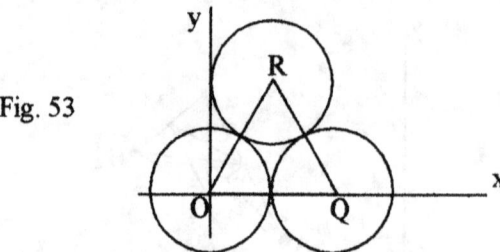

Fig. 53

83. In Fig. 54, find the coordinates of vertex U of △ STU if SU = 6 $\sqrt{10}$. The center T of the circle with radius 10 inches is at (6,6).

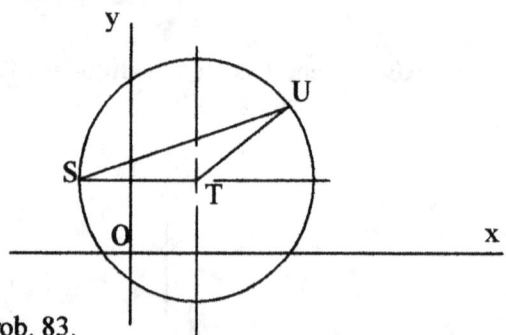

84. Calculate the area of △ STU in prob. 83.

85. The circle in Fig. 55 has its center at the origin and passes through point P as shown Show why, or why not, it would also pass through a second point S(17,14).

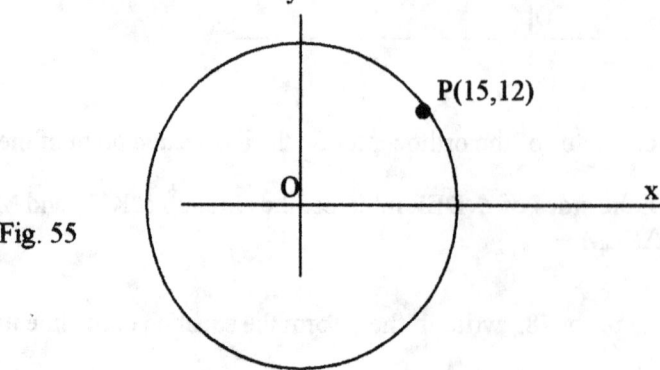

Fig. 55

86. The radius of a circle with center at the origin O(0,0) bisects a chord of the circle at S(4, 3). The length of the chord is 24 inches. Locate the intersection points M and N.

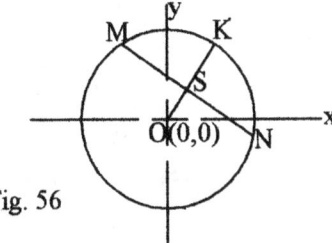

Fig. 56

87. Write the equation of the circle in probl. 86.

88. In prob. 86, solve for the coordinates of the intersection point K of the radius OK and the circle.

89. A circle has its center at C(h,k) and passes through Y(m,n). What is the equation of the circle?

Fig. 57

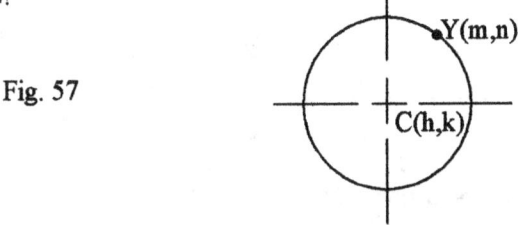

90. What is the equation of the circle with radius 6 inches, tangent to the x and y axes, has its center at C(h,k) and passes through P(a,b)?

Fig. 58

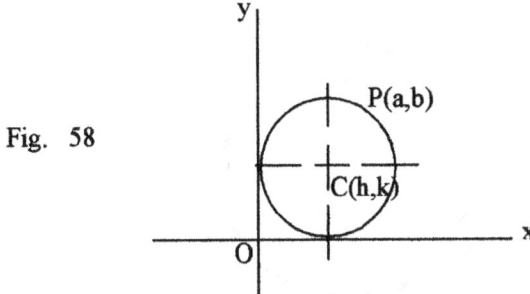

91. Each of the two congruent circles C and D has a radius of 4 inches, and their centers lie on the line m. Determine the equation of a line through the intersection point S that is perpendicular to m.

Fig. 59

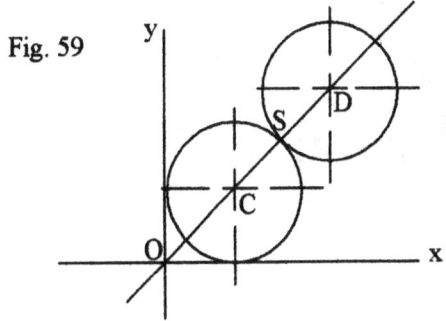

Areas of polygons

92. A regular pentagon is inscribed in a circle as shown in Fig. 60.

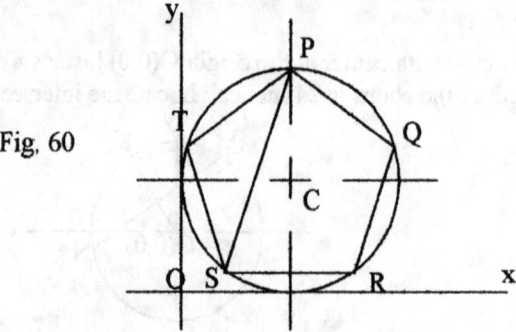

Fig. 60

The circle has a radius of 6 inches. Write the equation of SP.

93. Fig. 61 shows a regular pentagon ABCDE inscribed in a circle of radius 8 inches. Find its area.

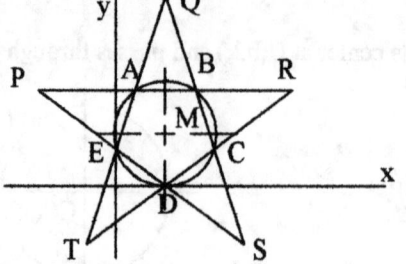

Fig. 61

93a. Calculate the area of the 5-pointed star in prob. 93.

94. The regular decagon in Fig. 62 is inscribed in a circle of radius 10 inches. Find the ratio of the area of the decagon to the area of the circle.

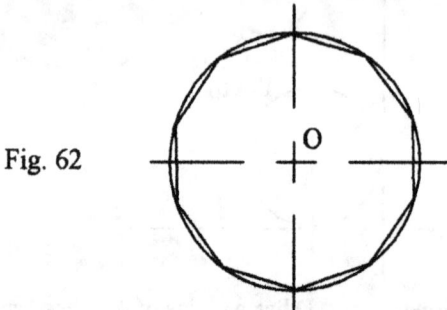

Fig. 62

95. Fig. 63 shows a regular polygon of n sides (n-gon) inscribed in a circle of radius r. If sides CD and FA of the n-gon are extended to intersect at T, express the measure of $\angle$T in terms of n and r.

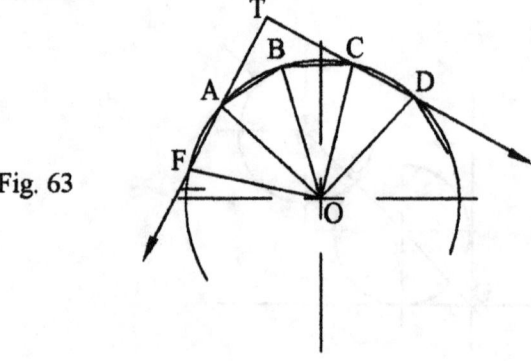

Fig. 63

96. Solve for the length, in terms of the side s, of the altitude of an equilateral triangle ABC with area (s ^{2} + 12s +36) $\sqrt{3}$.

Fig. 64

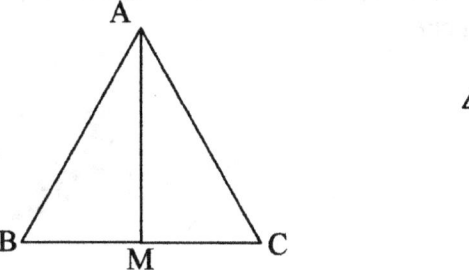

97. In the right triangle STU, the altitude TM is drawn to the hypotenuse. Find the area of △TUM. TU = 5 ft and ST = 12 ft.

Fig. 65

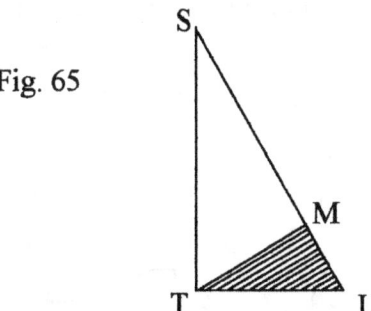

98. Each petal of the plastic sunflower model is a rhombus with diagonals AB and CD as shown in Fig 66 below.

Fig. 66

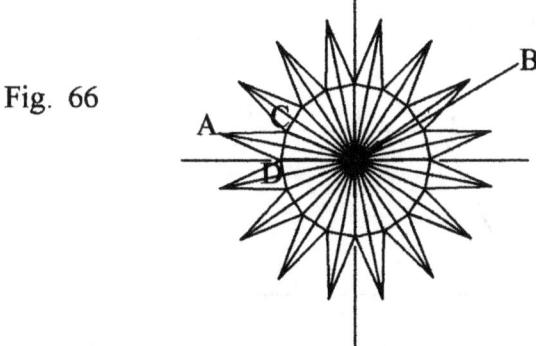

If the sunflower has 16 petals, show that the total petal area (assuming the area is flat) amounts to 8(AB)(CD).

99. Determine the area of the triangle CDE whose sides lie on the lines given below:

$$y = -3x/2 + 8$$
$$y = x/3 + 6$$
$$y = 11x/16 - 2$$

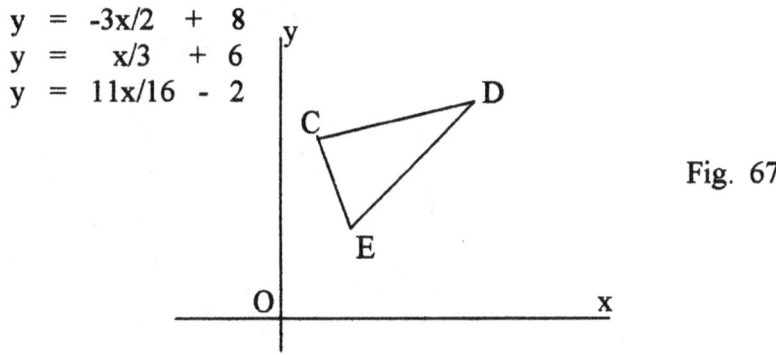

Fig. 67

100. Each leaf of this 3-leaved wildflower is made up of two segments of a circle facing each other. The segments are shaded as shown in Fig. 68. The radius of the circle is 10 inches.

Fig. 68

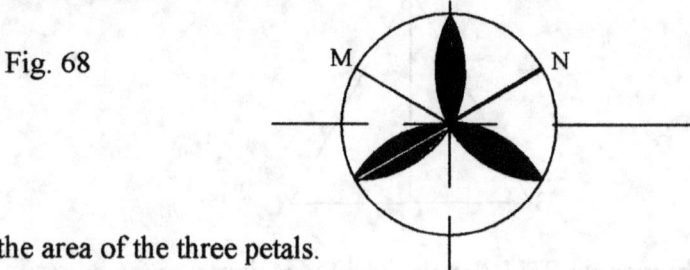

Find the area of the three petals.

101. A tiny island of 10 sq. mi. area has a circular park with a 0.5 mi. diameter. What is the probability that a parachutist jumping off a plane would land on the park?

Fig. 69

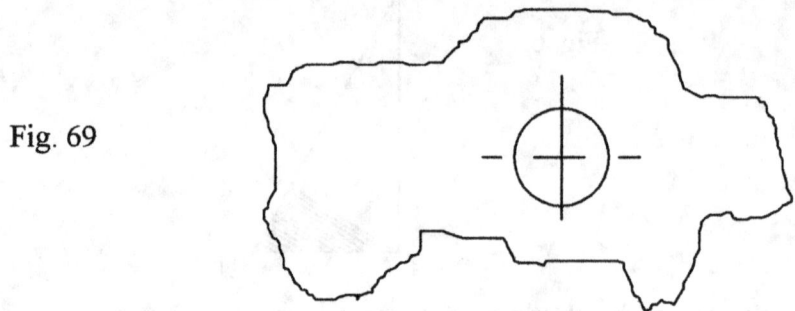

102. Derive, in two ways, the formula for the area A of a trapezoid, namely

$$A = 1/2 \, (a + b) \, h$$

Fig. 70

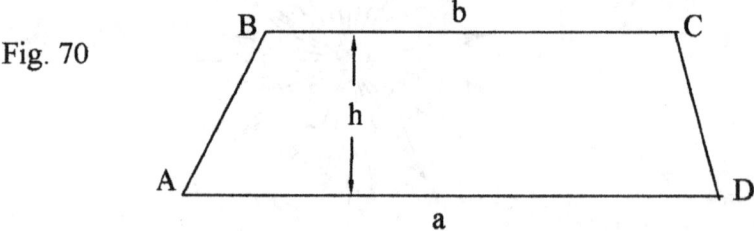

103. Find the total area of the eight segments of a circle formed by inscribing a regular octagon in a circle of radius 10 inches. What is the ratio of the area of the 8 segments to the area of the 8 sectors?

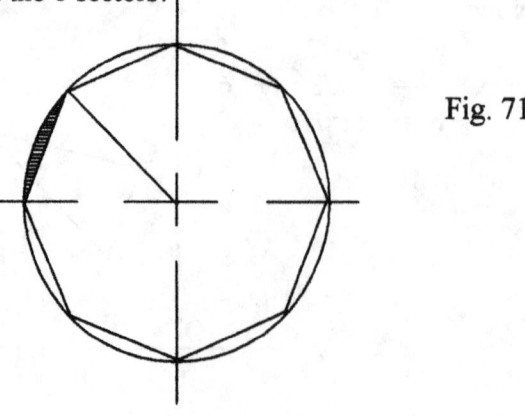

Fig. 71

Surface area and volume

104. Water from a cubical tank A is pumped out into a cylindrical container B that is 8 ft high. Find the minimum diameter of the cylindrical tank in order to contain the water from the cubical source. The cubical tank is filled with water and each edge of the cube is 8 ft long. Assume no loss in the transfer.

Fig. 72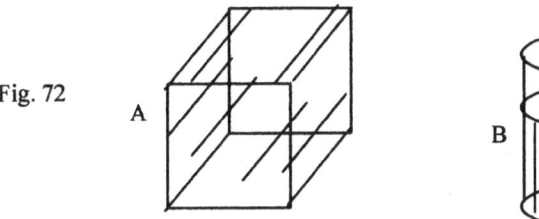

105. A solid piece of lead in the shape of a cube that is 10 inches on an edge is melted down into the shape of a cylinder with a base radius of 4 inches. How high would the cylinder be if no lead is lost in the melting process?

Fig. 73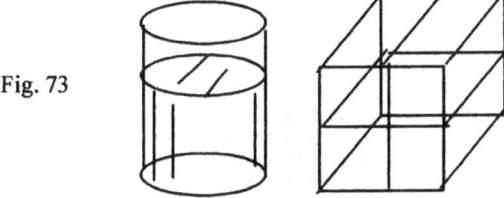

106. The base of a regular pyramid is a regular octagon with a radius of 12 in. and a slant height of 20 in. Solve for the volume of the frustum F of the pyramid formed by passing a cutting plane horizontally 3/4 of the way up across the altitude. See Fig. 74.

Fig. 74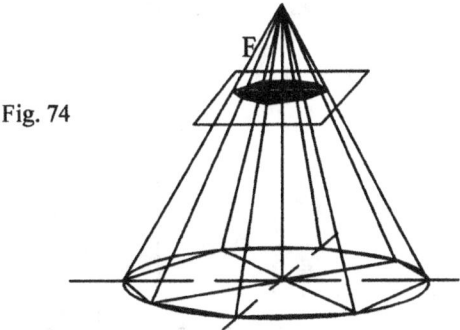

107. A regular triangular pyramid has an altitude of 24 in. and 26 in. slant height. Calculate the volume of the frustum F of the pyramid formed by passing a horizontal cutting plane 6 in. above the base, if the base measures $16 \sqrt{3}$ in. on a side.

Fig. 75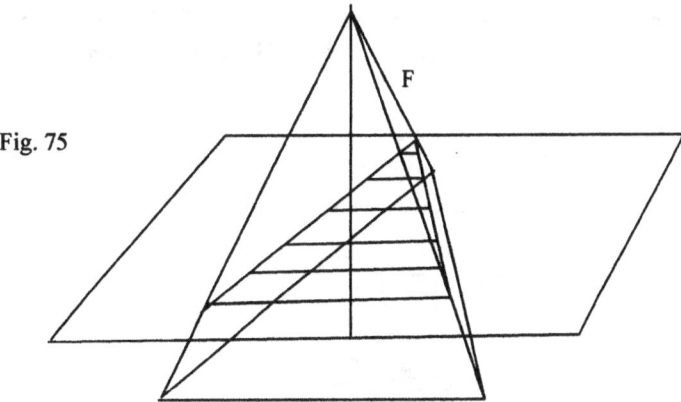

108. The radius of a cone is 6 in., its altitude 8 in. If it were made of solid lead, and then melted into a rectangular container with a 10 in. by 12 in. base, how high would the molten lead rise above the container's base?

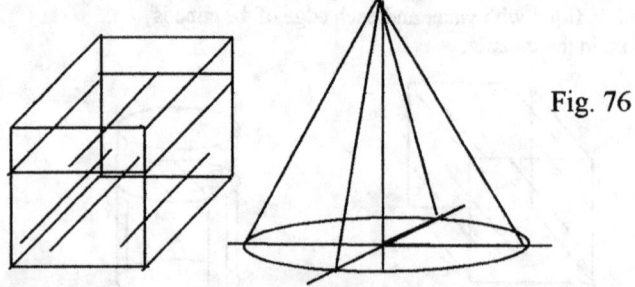

Fig. 76

109. A lead ball 20 in. in diameter is placed in a rectangular water tank that has dimensions 22 in. by 22 in. by 22 in.   If the tank is filled with water up to 22 in. high, how many inches of water will overflow when the ball is completely submerged in the water in the tank?

Fig. 77

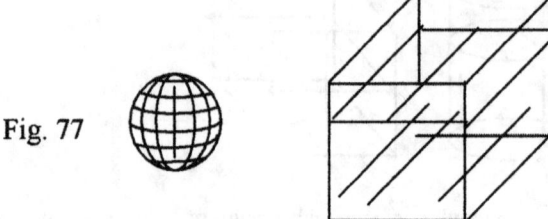

110. The great pyramid of Cheops has a 230 m x 230 m square base and an altitude of 145.69 m. Assuming that it has to be entirely covered with canvas, how many sq. m. of the material would be needed, and how many cubic feet of concrete would be needed to duplicate such a towering structure?

Fig. 78

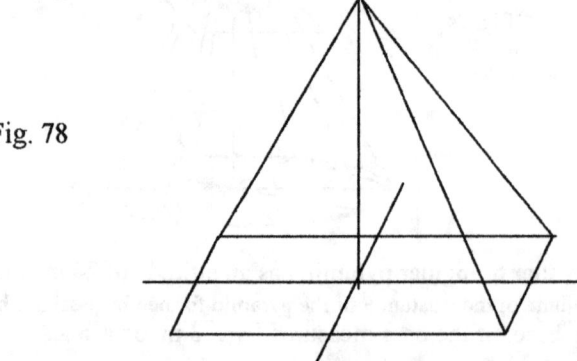

111. A sphere of diameter 12 in. is inscribed in a cylinder. The sphere is made of non-porous material and does not absorb liquids. How much water can be poured into the cylinder before overflow occurs?

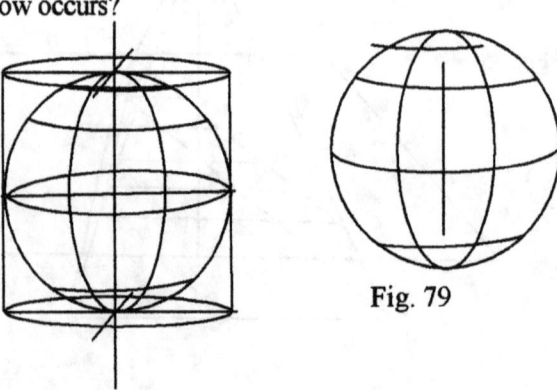

Fig. 79

Constructions and Loci

112. Given, the segment KL (Fig. 80). Divide KL into three segments KA, AB, and BL such that AB is twice the length of KA and BL is twice the length of AB.

Fig. 80   K_____ L

113. Fig. 81 shows two angles, ∠ M and ∠ N , with magnitudes of m and n degrees, respectively.

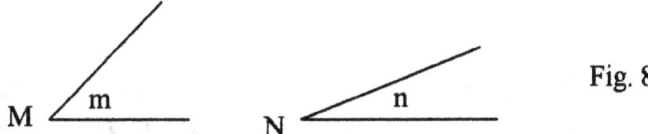

Fig. 81

Construct an angle with a measure of (180 - 2m + n) / 2 degrees.

114. Referring to Fig. 81 in prob. 113, construct an angle with a measure of [180 - (m - n)] / 2 degrees.

115. Given, △KLM (Fig. 82)

Fig. 82

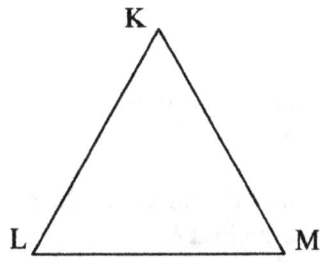

Construct a segment with a length equal to   { [(1/3 KL) (2/3 KM)] / [(2/3 KL)] }
$$\frac{\{ [(1/3\ KL)(2/3\ KM)] / [(2/3\ KL)] \}}{3}$$

Hint: Make (2/3 KL) and (1/3KL) the two segments of the left side of the triangle, the two segments of the right side of the triangle would be (2/3 KM) and a segment of length x such that

(2/3 KL) / (1/3 KL)   =   (2/3 KM) / x

Then construct a segment of length 1/3 x. This is the desired segment.

116. Referring to Fig. 82 in prob. 115, construct a segment with a length equal to

$$\frac{[(2/3\ KL) (1/3\ KM)] \ /\ (1/3\ KL)]}{3}$$

117. In prob. 83, locate a point M on side AB of   ABC such that BM/MA = 3/4, and then locate a point N on side BC so that BN/ NC = BN/MA.

Fig. 83

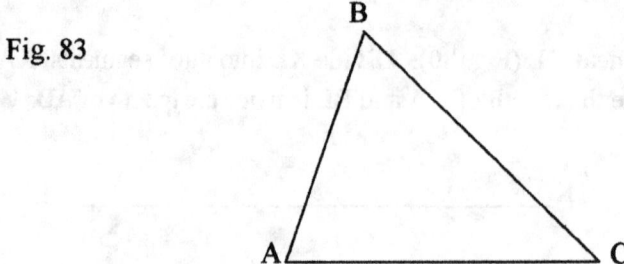

What are the coordinates of M and N?

118. In Fig. 84, circle O has its center at the origin, and diameter KL is inclined at 45 degrees to the y-axis. Through the origin, draw another diameter MN with MN ⊥KL.

Fig. 84

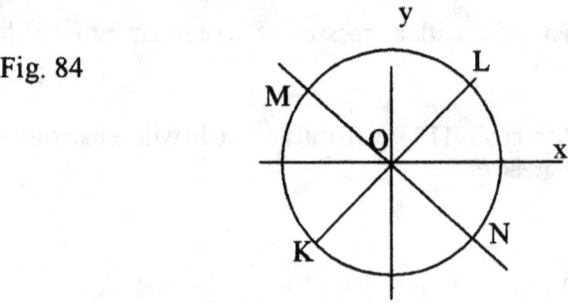

119. In Fig. 84, prob. 118, determine the equation of the tangent to the circle with M as the point of tangency.

120. In Fig. 84, prob. 118, solve for the equation of the line through L perpendicular to the tangent to the circle at N.

121. Given the segments PQ and RS shown in Fig. 85:

Fig. 85   P——————————— Q
                 R ——————————————— S

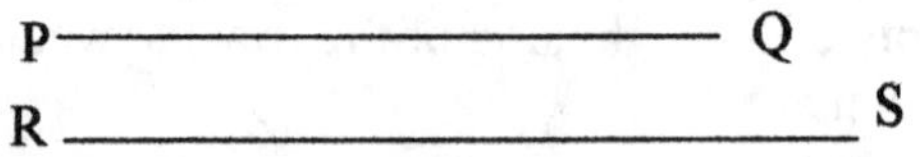

Construct a segment with a length equal to $\sqrt{2 (PQ) (RS)}$.

122. Given, the segments p and q in Fig. 86.

Fig. 86 $\dfrac{p}{q}$

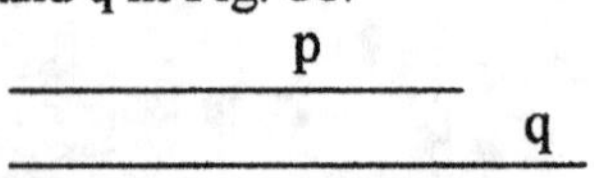

Construct a segment r such that $r = 3\sqrt{pq}$.

123. In Fig. 86, prob. 122, construct a segment s such that $s = \sqrt{5/2\ pq}$.

124. Given, the segments AB and PQ in Fig. 87. Construct a segment k with a length equal to the mean proportional between 2AB and 3PQ.

A ——————————— B

Fig. 87   P ————————————— Q

125. Referring to prob. 124, use the given segments AB and PQ to construct a regular octagon with a perimeter of 2AB + 3PQ.

126. The right triangle SOT in Fig. 88 has its vertices at S(5,0), T(0,12) and O(0,0). Locate its centroid.

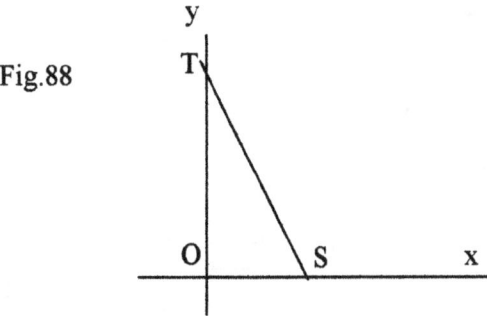

Fig.88

127. Referring to Fig. 88 in prob. 126, connect the centroid to the origin, and write the equation of the line through these points.

128. Referring to Fig. 88 in prob. 126, locate the circumcenter.

129. Referring to Fig. 88 in prob. 126, connect the circumcenter to the origin, and then write the equation of the line through these points.

130. Referring to Fig. 88 in prob. 126, locate its orthocenter.

131. Referring to prob. 130, connect the orthocenter to the origin, and then write the equation of the line through these points.

132. Connect the three points that have been located, namely the centroid, circumcenter, and the orthocenter, to obtain the classic line known as Euler's Line. Write the eqution of this line.

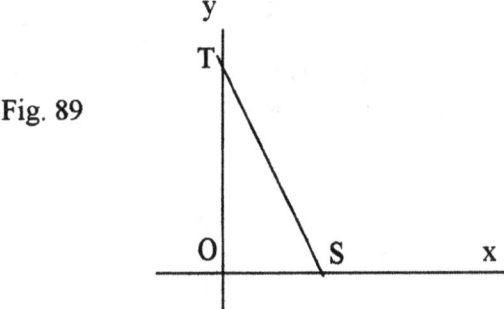

Fig. 89

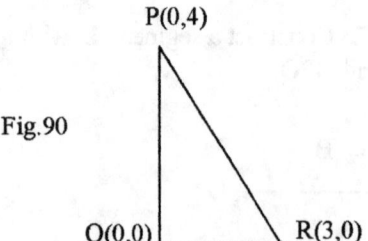

Fig.90

133. Given, the right triangle POR in Fig. 90. Locate its centroid and determine its distance from O. If the dimensions of the triangle are doubled, by what percent is the distance OC of the centroid C from O increased?

134, Find the locus of points in a leg and hypotenuse of a right triangle as it rotates about the perpendicular bisector of its hypotenuse.

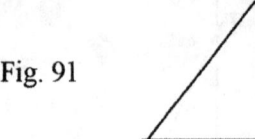

Fig. 91

135. What is the locus of points in a sphere of radius 3 in. as the sphere rolls (its center describing a circle) inside a cylinder of radius 10 in. ?

Fig. 92

134. Find the locus of points in a leg and the hypotenuse of a right triangle as the triangl rotates about the perpendicular bisector of its hypotenuse.

136. Describe the locus of the intersection of tangents to a circle drawn at the endpoints of 10 equally-spaced diameters.

Fig. 93

137. Determine the locus of points in a sphere of radius 3 in. as the sphere rolls straight ahead on a horizontal plane.

Fig. 94

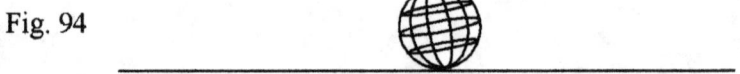

138. Describe the locus of points equidistant from A, B, C, D, and E in the regular pentagon in Fig. 95, and equidistant from A and B.

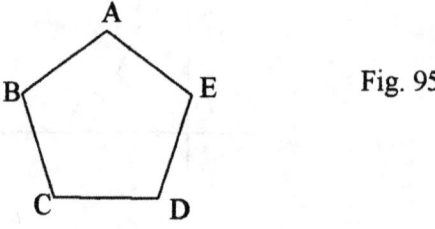

Fig. 95

·139. Given, the regular hexagon in Fig. 96. Describe the locus of points equidistant from P and Q.

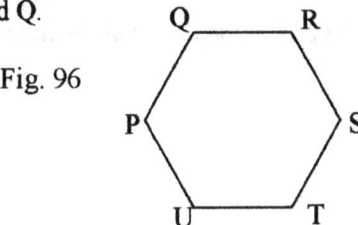

Fig. 96

140. In the preceding problem, describe the locus of points equidistant from R and U.

141. In Fig. 96, prob. 139, what is the locus of points equidistant from R, S, and T?

142. Referring to Fig. 96, prob. 139, describe the intersection of the locus of points equidistant from S and T, and equidistant from U and R.

143. Points O(0,0), B(6,0), and C(10,8) lie on the x-y coordinate plane. What is the locus of points equidistant from O, B, and C, and determine the coordinates of the point of intersection of the locus and the x-y plane.

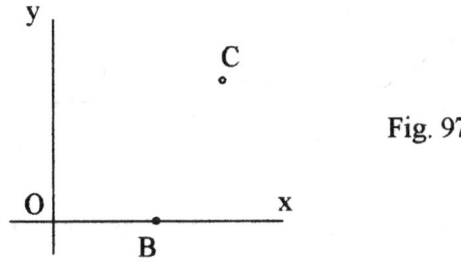

Fig. 97

144. Construct an ellipse with the major and minor diameters of 8 in. and 5 in. respectively.

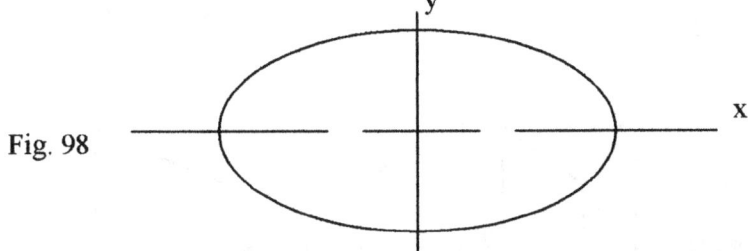

Fig. 98

145. Given, the major and minor diameters of 7 in. and 5 in. respectively, of an ellipse. Construct the ellipse.

146. Construct a parabola with directrix x = 8, and the x-axis as the axis of symmetry. The focus is at x = -6.

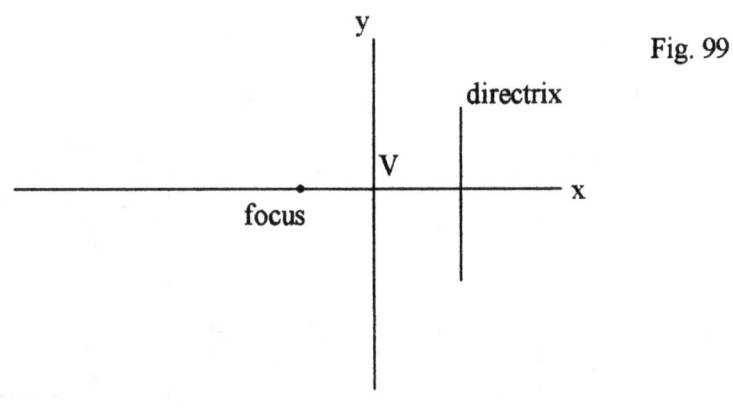

Fig. 99

Introduction to Descriptive Geometry

147. Draw the missing top view and the true size projection (TSP) of the inclined surface in the front view (left-side drawing in Fig. 100).

Fig. 100

Front view   Right side view

148. Fig. 101 shows the front view of a rectangular prism with an inclined upper base. Draw the missing top view and the TSP of the inclined surface in the front view.

Fig. 101

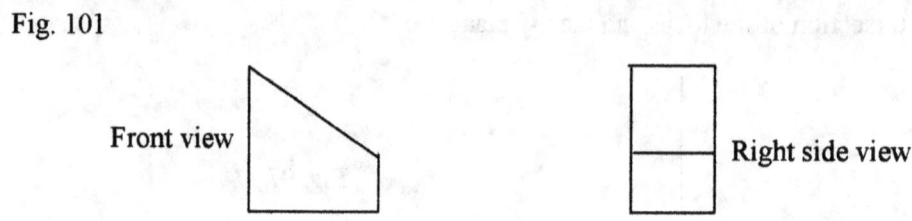

Front view   Right side view

149. Draw the missing top view and the TSP of the inclined upper base of the triangular prism in Fig. 102.

Fig. 102

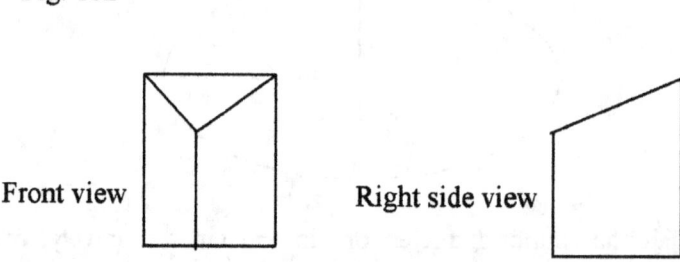

Front view   Right side view

150. Draw the missing top view and the TSP of the inclined upper base of the triangular prism in Fig. 103.

Fig. 103

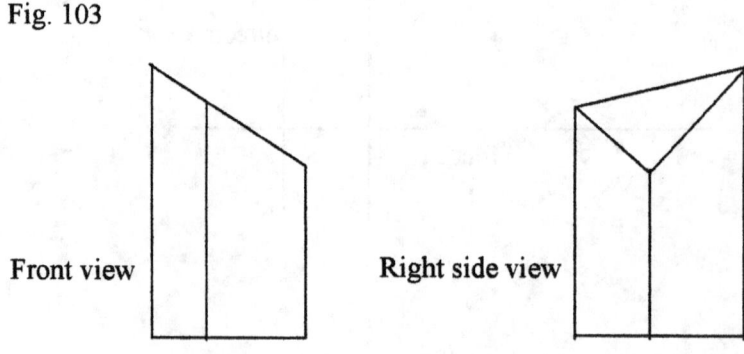

Front view   Right side view

Geometric transformations

151. Given: ABCDE ———→ PQRST
The line of reflection is the line y = x. In domain D are A(0,5), B(-5,5), C(-9,3), D(-9,-1), and E(-7,-3). Find the distances of A, B, C, D, and E to the line of reflection.

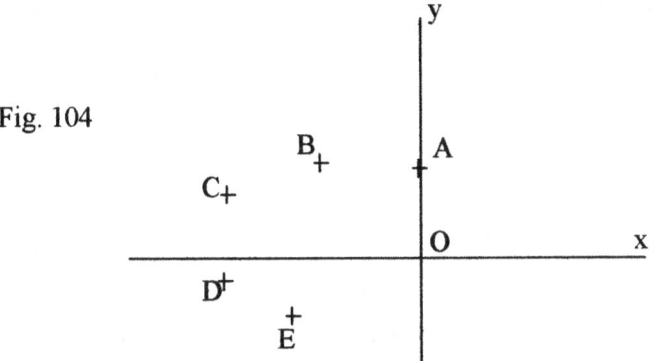

Fig. 104

152. Referring to Fig.104, prob. 151, solve for the equation of the line perpendicular to the line of reflection and passing through the midpoint of BC.

153. In prob. 151, determine the coordinates of P, Q, R, S, and T.

154. Referring to Fig. 104, prob. 151, calculate the slope of the line through B and T.

155. Find the equation of the translational image of the line y = -2x + 2, if the translation maps (x,y) into (x+6, y+10).

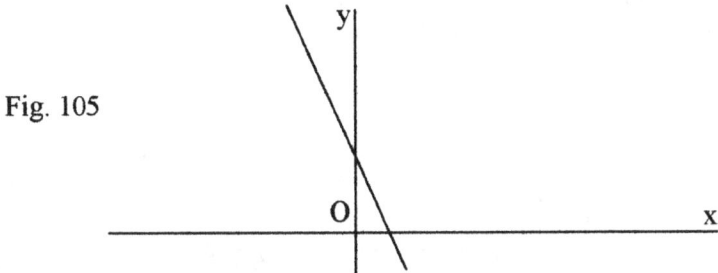

Fig. 105

156. Referring to prob. 155, solve for the y-intercept of the translational image of the line y = 4x - 2 under the given translation.

157. Consider AB as the wiper blade of a car and solve for the area brushed by the blade through 120 degrees of rotation. The center of rotation is at the intersection point of the y-axis and the line AB in Fig. 106.

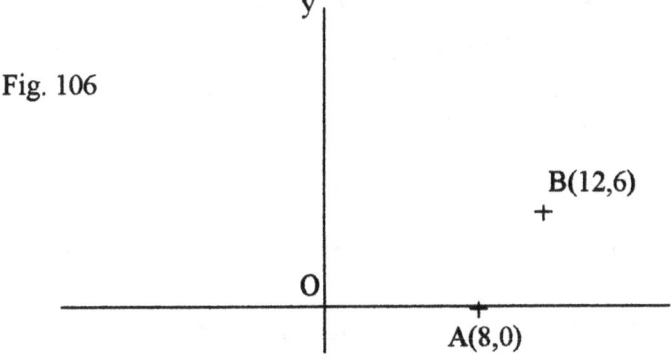

Fig. 106

158. Given, the triangle ABC with vertices at A(10,0), B(18,0) and C(14,6). Draw the rotational image under a rotation of 270 degrees with the origin as the center of rotation, and locate the image points A', B', and C'. Determine the arc length, in linear units, of arc AA'. (Fig. 107)

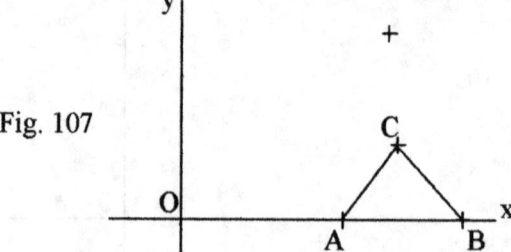

Fig. 107

159. Referring to prob. 158, write the equation of the line from B to its image point B'.

160. In prob. 158, solve for the area of the sector bounded by BOB'.

161. In prob. 158, determine the area of the sector bounded by AOA'.

162. In prob. 158, calculate the area of the triangle AA'C' formed by the rotation.

163. Referring to prob. 158, calculate the length, in linear units, of arc CC'.

164. Fig. 108 shows a cube under a transformational dilation with k = 3. Determine the coordinates of the image points K', L', and M'.

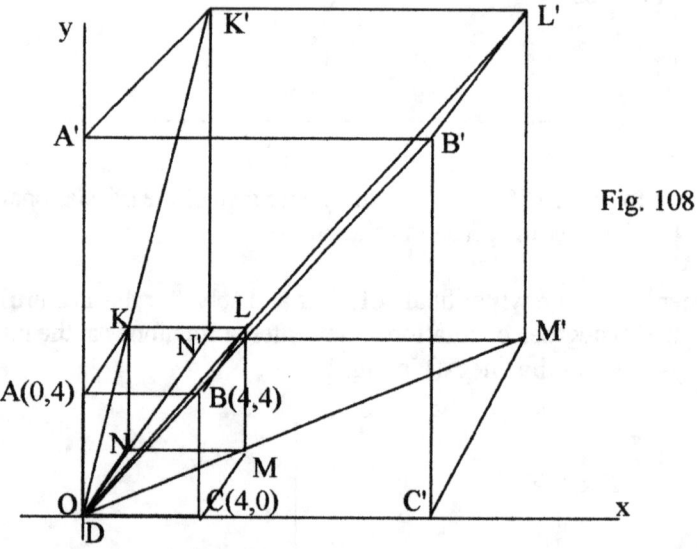

Fig. 108

165. In prob. 164, solve for the length of the diagonal of the image cube.

166. Referring to prob. 164, solve for the volume of a sphere circumscribed about the image cube.

167. In prob. 164, determine the surface area of a sphere inscribed in the image cube.

V e c t o r s

168. A speedboat headed in a northerly direction is acted upon by a gale force of 200 fps N42E. What is the useful component of the wind force along the northerly direction?

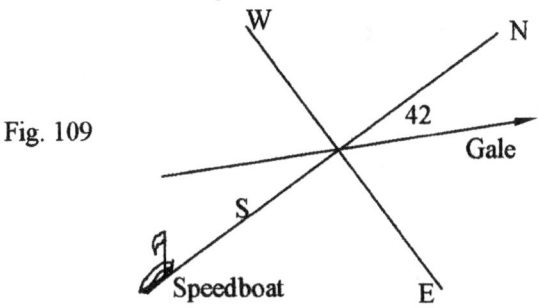

Fig. 109

169. If the boat in prob. 168 is speeding at 100 fps, and the wind blowing at 30 fps, find the boat's resultant direction and velocity.

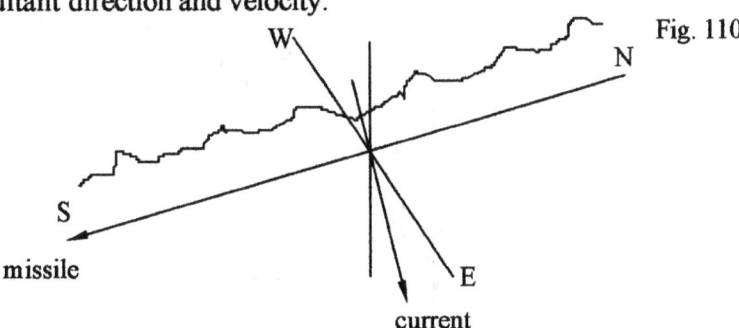

Fig. 110

170. An underwater missile moves at 100 fps in still water. If such a missile is fired horizontally southward in water where the current moves at 30 fps S42E, solve for the direction and magnitude of the resultant velocity of the missile. (See Fig. 110)

171. A submarine is cruising at a speed of 50 knots in still water. While submerged and moving horizontally southward, it encounters a vertical current at 15 knots. Find the magnitude and direction of the submarine's resultant velocity.

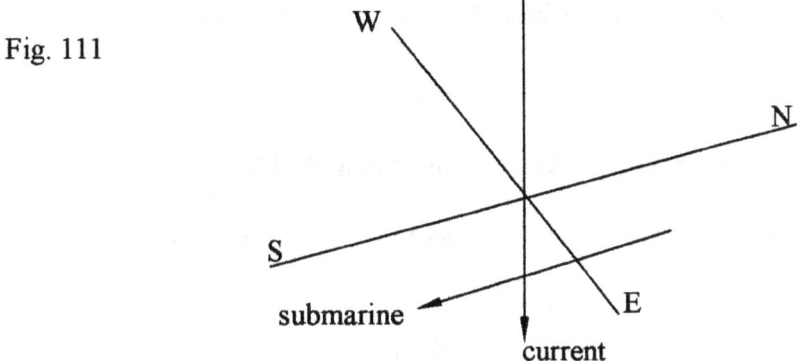

Fig. 111

172. A missile from a gun emplacement speeds through the atmosphere's still air at 1500 mph in a northerly direction. If the trajectory is 50 degrees above the horizontal above a blowing wind at 15 knots eastward, how fast would a second missile fired 3 seconds afterward from the same emplacement, travel in the same trajectory as that of the first missile so that both missiles would score direct hits simultaneously on a target 100 miles away. Assume zero effect of gravity.

Introduction to unit vectors

173. Find the magnitude of each vector:
     P = 2i + 6j + 4k
     Q = 4i + 6j + 4l

174. Determine the magnitude of each vector:
     R = 4i + 6j + 4k
     S = 6i + 4j + 4k

175. Calculate the magnitude of each vector:
     K = 2i + 6j + 4k
     L = 6i + 6j + 4k

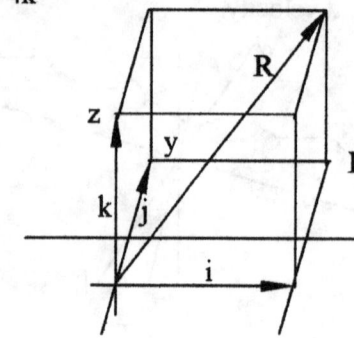

Fig. 112   Unit vectors

176. In probs. 173 and 174, find the value of P + Q + R.

177. In prob. 173, determine the value of P - Q.

178. In probs. 173 and 174, what is the value of R - P ?

179. Find the direction cosines of the vectors in prob. 174.

180. What are the direction cosines of the vectors in prob. 175?

181. Determine the direction cosines of the vectors in prob. 173.

182. Find the angle between vectors P and Q in prob. 173.

183. Determine the angle between vectors R and S in prob. 174.

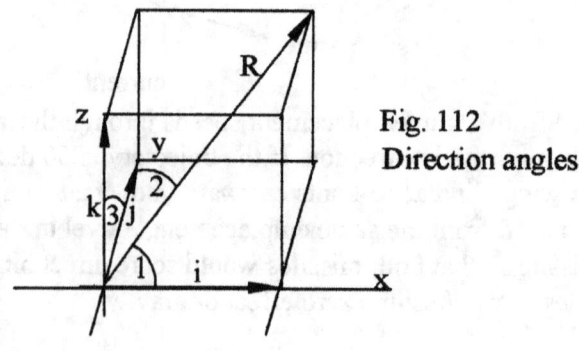

Fig. 112
Direction angles

184. What is the measure of the angle between K & L in prob. 175?

185. Given: K = 2i + 4j + 0k   and   L = 6i + 4j + 4k.
Find the scalar or dot product of K and L.

186. Find the scalar product of: P = 2i - 3j + k,   Q = -i + 2j - k.

187. Find the cross or vector product of K = 4i - 6j - 6k and L = 8i + 4j + 2k.

.88. What is the cross product, K x L, of K and L in prob. 187?

189. Given: P = 2i - 3j + k,   Q = -i + 2j - k, and R = 0i + 0j + 4k.
Find the value of P · (Q x R).

190. Given, the vectors P, Q, and R in prob. 189. Determine the value of
(P · Q) x R.

Introduction to basic engineering mechanics (optional)
for enrichment

191. (Fig. 114). A weight of 1000 lbs. hangs at the end of a 14 ft uniform
beam as shown. The cord T makes an angle of 45 degrees with the beam.
Find the magnitude of the force in T.

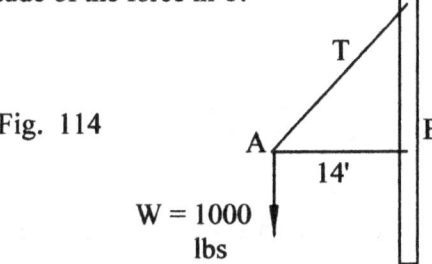

Fig. 114

192. Determine how much compression force there is on AB in prob. 191.

193. Given, the vector system in Fig. 114. If the suspended weight W is
100 lbs, determine the tension forces on OA and OB.

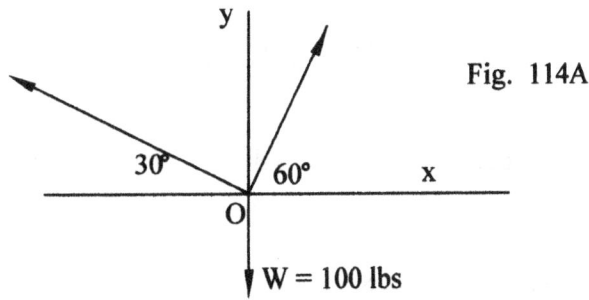

Fig. 114A

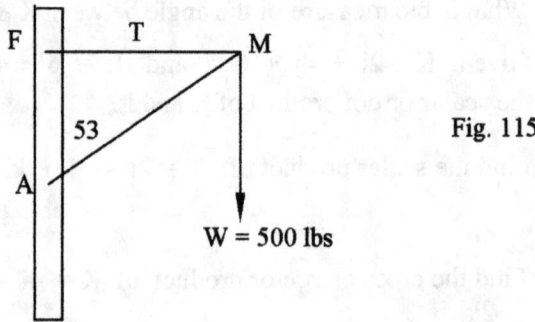

Fig. 115

194. The system shown in Fig. 115 is in equilibrium. The uniform boom C is 3 ft long. Find the magnitude of T and the compression C on the boom, if W = 100 lbs, and FM is perpendicular to FA,

Introduction to complex numbers (Optional)
for enrichment

195. Convert from Cartesian to polar form the complex number - 9 - i6.

196. Given, the complex number 50 + i40. Convert the number to polar form.

197. The complex number 10 - i4 is in Cartesian form. Change it to polar form.

198. Convert from polar to Cartesian form the complex number 7 $\angle 120°$.

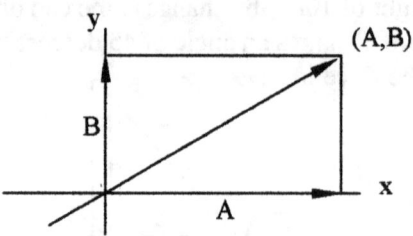

Fig.116   Cartesian coordinates of a complex number

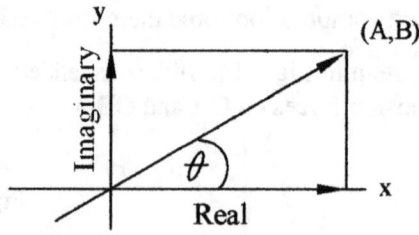

Fig. 117   Polar coordinates of a complex number

199. Determine the roots of x^2 +   6x + 12 = 0. Express them in both Cartesian and polar forms.

Multiply the following complex numbers:

200. (3 + i6) x (10 - i7)

201. (-6 - i5) x (8 + i8)

202. (10 - i2) x (-16 + i7)

Divide the following complex numbers:

203. (3 + i6) ÷ (10 - i7)

204. (-6 - i5) ÷ (8 + i8)

205. (10 - i2) ÷ (-16 + i7)

Determine the complex conjugate of the following complex numbers:

206. (7 + i4) *

207. (-12 - i15) *

208. (8 - i8) *

EULER'S LINE

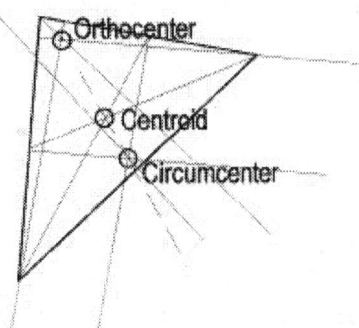

Orthocenter

Centroid

Circumcenter

EULER'S LINE

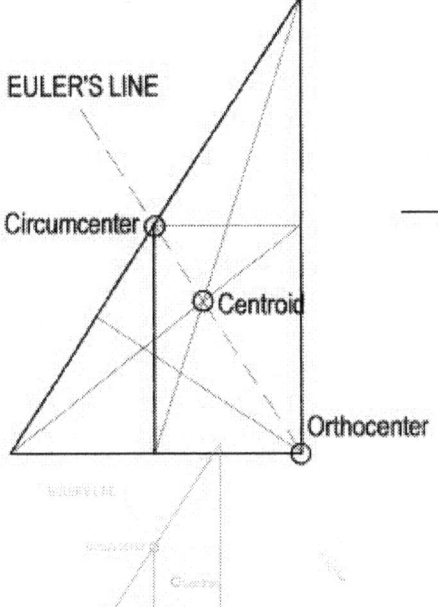

Circumcenter

Centroid

Orthocenter

Part Two

Answers

1. (See Fig. 201A and 201B)

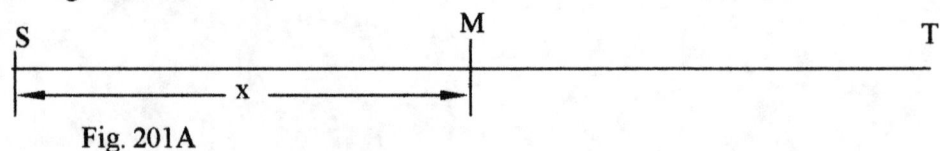

Fig. 201A

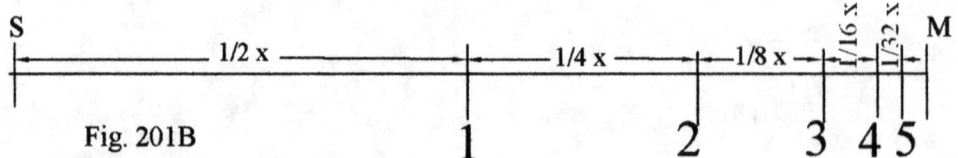

Fig. 201B

Let M be the halfway point from S to T, and x be the halfway distance from S to T. According to the problem, runner S would run to point M, return halfway back, and therefore, at the end of trip 1 would be at point 1 (Fig. 201B);
At the end of trip 2, he would be at pt 2;
At the end of trip 3, he would be at pt 3;
At the end of trip 4, he would be at pt 4; and
At the end of trip 5, he would be at pt 5.

Distance of point 5 to M $= 1000 - (1/2 x + 1/4 x + 1/8 x + 1/16 x + 1/32 x)$
$\qquad\qquad\qquad\quad = 1000 - 968.75$, since $x = 1000$
The distance between the runners would then be $2 \times 31.25 = 62.5$ ft.

2. Distance D traveled by S = the distance traveled by T = the sum of $d_1 + d_2 + d_3 + d + d_5$ below:

$\qquad d_1 = x + 1/2 x = 3/2 x = 1500$
$\qquad d_2 = 1/2 x + 1/4 x = 3/4 x = 750$
$\qquad d_3 = 1/4 x + 1/8 x = 3/8 x = 375$
$\qquad d_4 = 1/8 x + 1/16 x = 3/16 x = 187.5$
$\qquad d_5 = 1/16 x + 1/32 x = 3/32 x = 93.75$
Hence D = 2906.25 ft

3.

| B | | | D | E F C | | A |

-52   -30   -8

BC $= 1/2 (^-8 - ^-52) = 22$
coordinate of C is $-52 + 22 = -30$

BD $= 1/3 (^-8 - ^-52) = 14.67$
coordinate of D is $-52 + 14.67 = -37.33$

DE $= 1/4 BD = 1/4 (14.67) = 3/67$
coordinate of E is $-37.33 + 3.67 = -33.66$

EF $= 1/5 DE = 1/5 (3.67) = .73$
coordinate of F is $-33.66 + .73 = -32.93$

Coordinates of A and F are -8 and -32.93 respectively.

The distance from A to F is $^-8 - ^-32.93 = 24.93$.

4.

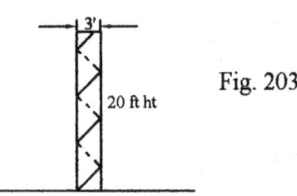

Fig. 203

The cylinder in Fig. 203 may be cut vertically downward and then spread out to assume the shape of a rectangle as shown below:

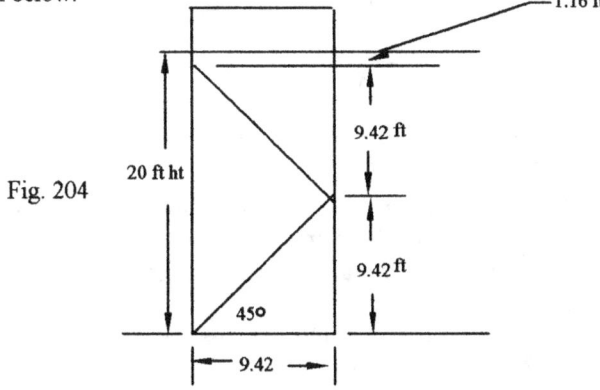

Fig. 204

The line climbs at an angle of 45 degrees, so its rise is equal to the horizontal distance it travels. The width of the rectangle is equal to the circumference of the cylinder which is ~d, or 3.1416 x 3 = 9.42 ft. While the line travels a distance of 9.42 ft horizontally, it also rises by 9.42 ft. Therefore, the number n of turns the line makes to reach the top of the cylinder = 20 /9.42 = 2.12 turns.

5. It is given that the line k does not intersect plane P, which means that k lines on P as shown in the figure below.

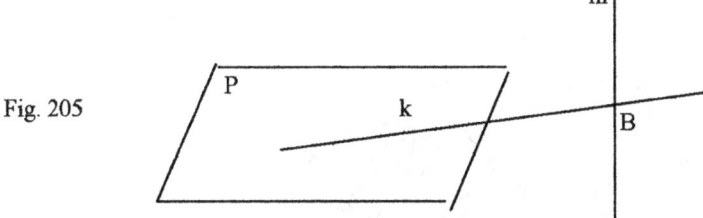

Fig. 205

From the figure above, it is clear that if both k and P intersect a line m at some point B, B has to lie on the plane since k and P are coplanar. Note also that a plane has infinite length and width, hence any intersection point of line k and any other line would be a point on the plane P.

6. The figure below shows two pentagonal pyramids, one base against the other, and each having five triangular sides - each triangular side a plane. There are 10 planes in all. The 15 intersections are: VA, VB, VC, VD, VE, AB, BC, CD, DE, EA, YA, YB, YC, YD, and YE.

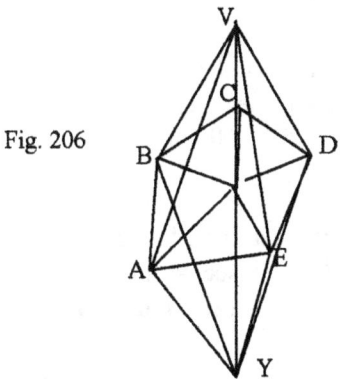

Fig. 206

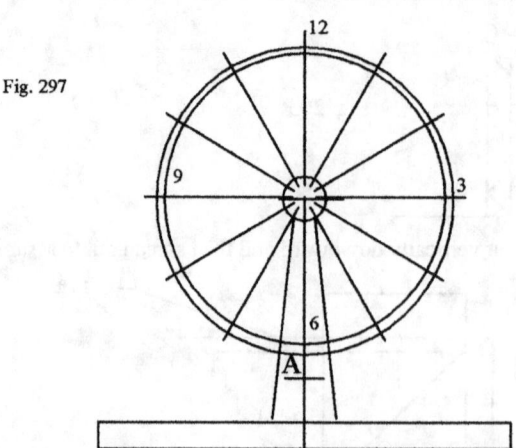

Fig. 297

Fig. 207 shows a Ferris Wheel with its 12 radial beams numbered as shown. There are 12 beams around 360 degrees, hence there are 30 degrees between beams 5 and 6. The wheel makes one turn every 1/2 minute, or 2 revolutions per minute (rpm).

Time for beams 5 and 6 to pass through point A = angle between beams / turning speed of wheel = 30 / 720 = 1/24 min or 2.5 sec.

B. If the wheel in prob. 7 has 14 beams instead of 12, then the angle between successive beams is 360 / 14 = 25.71 deg.

The angle between beams 5 and 12 = (12 - 5) 25.71 = 180 deg.
The angle between beams 1 and 6 = (6 - 1) 25.71 = 128.55 deg.
The difference between the two angles = 180 - 128.55 = 51.35 deg
0r 51.45 / 57.29 = .90 radian.

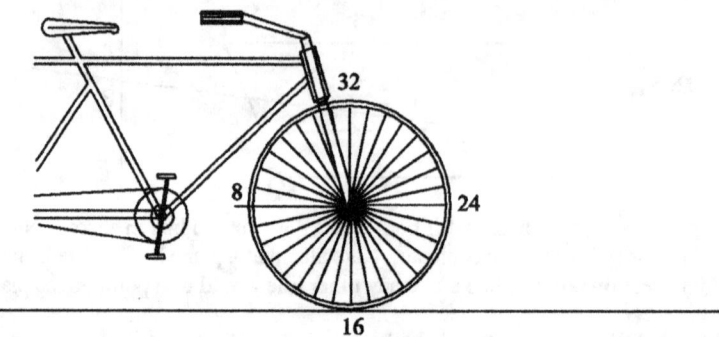

9. The angle between two successive spokes in the bicycle wheel = 360 / 32 = 11.25 deg. The angle between spokes 11 and 27 = (27 - 11) 11.25 = 180 deg.

10. Turning speed of the bicycle wheel = 10 x 11.25 or 112.5 deg per 1/2 sec or 225 deg / sec. Since a minute = 60 sec , the 225 deg / sec speed = 225 x 60 deg / min = (225 x 60) / 360 or 37.5 turns per min.

11. There are 8 angles between 3 consecutive white stokes, hence the speed required for 3 white spokes to pass a point A in 0.9 sec = no. of deg between 3 white spokes / time = [8 x (360 / 14) / 0.9] = 228.57 deg / sec.

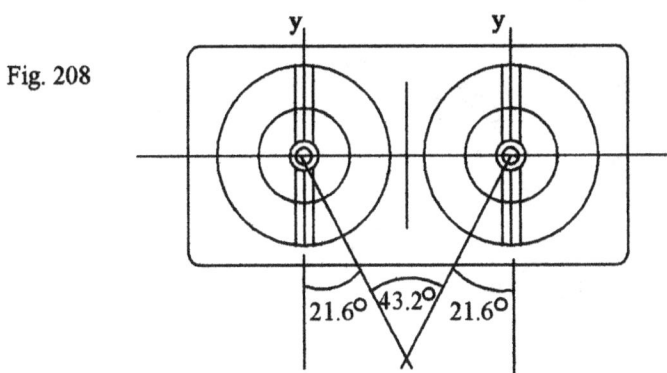

Fig. 208

12. Blade turning speed is given as 540 rpm or 9 rps

At the end of 1/16 sec, each blade has traveled 1/16 (9) or .56 revolution. A .56 turn, with the positive y-axis as the starting point is equal to 201.6 deg. (See Fig. 208) Therefore, the angle between the blades is 43.2 deg.

13. For the blades to form a 90 deg angle, each must travel 225 deg (or .63 turn) past the starting point.
Since angle traveled = turning speed x time, the desired time t would be
t = angle traveled / speed = .63 / 9 = .07 sec.

14. Speed of the hour hand = 5 divisions per hour or 30 deg / 60 min = 0.5 deg / min.
The time shown is 11:20:15 which is 20.25 min. past 11:00 (see Fig. 209).
Hence, the hour hans traveled an angle equal to 0.5 (20.25) = 10.13 deg.

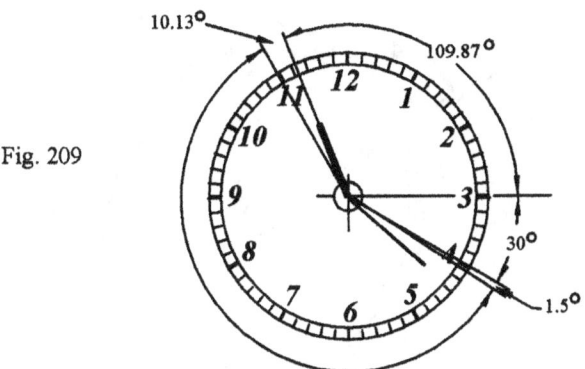

Fig. 209

Angle traveled by min hand = min hand speed x time of travel.
Speed of min hand = 360 deg / hr = 0.1 deg /sec; Time of travel = 15 sec
Therefore, angle traveled = 0.1 (15) = 1.5 deg

Speed of the hr hand = 0.5 deg/ min; in 15 sec it travels .13 deg.
Hence the angle between the hr hand and sec hand = (30 - .13) + 90
= 119. 87 deg
The angle between the sec hand and the min hand = 30 + 1.5 = 31.5 deg.

15. From prob. 14, we see the speeds of the hr and min hands of the clock as:
Speed of hr hand = 0.5 deg / min = .0083 deg / sec; and
Speed of sec hand = 6 deg / min = 0.1 deg / sec.

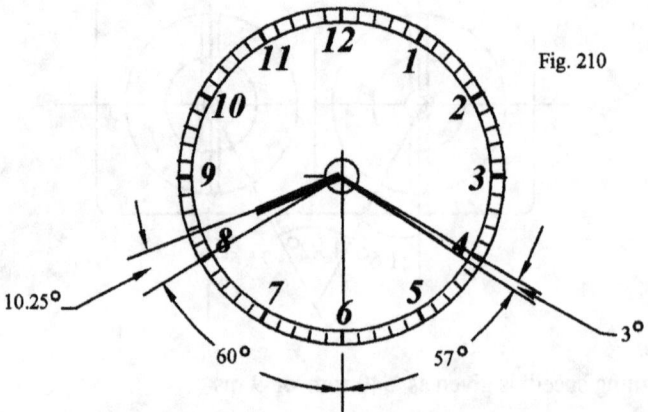

Fig. 210

15. From the preceding problem, speed of the hr hand - 0.5 deg / min, and speed of the min hand = 6 deg / min. In Fig. 210, the angle traveled by the hr hand in 20.5 min = 0.5 (20.5) = 10.25 deg. The angle traveled by the min hand past 8:20 = 6 (0.5) = 3 deg. Hence, the angle between the hr and sec hand = 60 + 10.25 = 70.25 deg; and the angle between the min and sec hand = 60 - 3 = 57 deg.

16. Referring to Fig. 211, for the min and hr hands to form a right angle, the min hand should be ahead of the hr hand by 15 divisions.
At 10 o'clock, the min hand's position is such that it is ahead of the hr hand already by 15 divisions, there need be only 5 more.

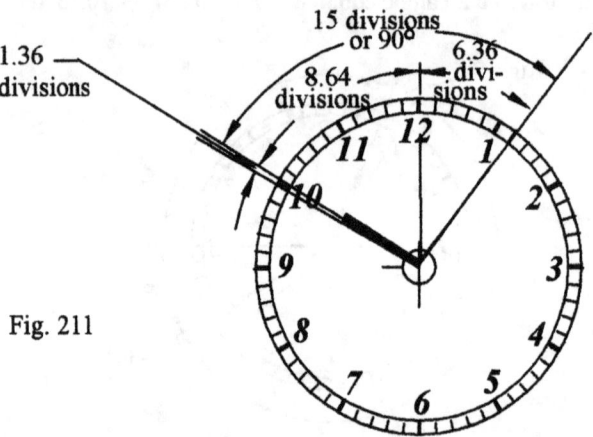

Fig. 211

In the period of one hour, the min hand travels 60 divisions, the hr hand only 5 divisions, hence the difference in the travels of both in one hour is 55 divisions. In how much time would a difference of 15 divisions occur? The answer is given by the simple proportion, namely:

$$55 \text{ divisions } / 60 = 15 \text{ divisions } / x$$
$$x = 16.36 \text{ divisions or minutes}$$

But since the min hand is already ahead of the hr hand by 10 divisions or minutes, take off 10 from 16.36 divisions to make it just 6.36.

Therefore, the two hands would be at right angles at exactly 10:06:36. (Figure out another solution by using 45 divisions instead of 15).

17. The earth rotates about its axis once every 24 hours.
Degrees traveled in 24 hours of rotation = 360 degrees.
Rotation time = 3 hrs, 12 mins = 3.2 hrs = 3.2 / 24 (360) = 48deg
Radians traveled in 3.2 hrs = 48 / 57.29 = .84 radian

18. The mean distance of the sun to the earth is 90,000,000 miles. Its period is 365.25 days; that is, in 365.25 days, the earth revolves once around the sun.
Therefore, the earth rotational speed = 360 . (365.26) = .04 deg / hr
Degrees traveled in 3.2 hrs = .04 (3.2) = .13 deg
Radians traveled in 3.2 hrs = .13 / 57.29 = .002 radian

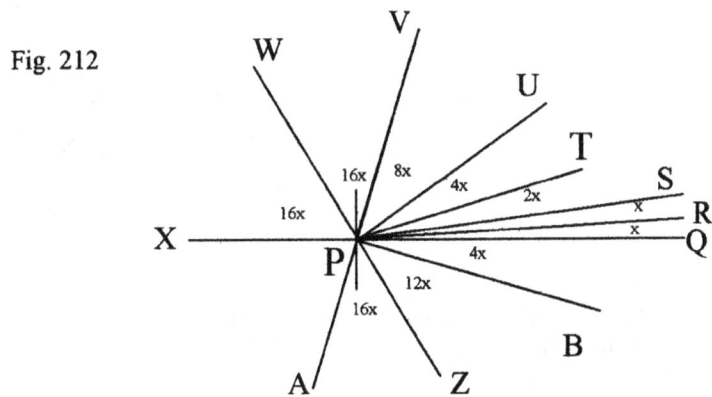

Fig. 212

20.

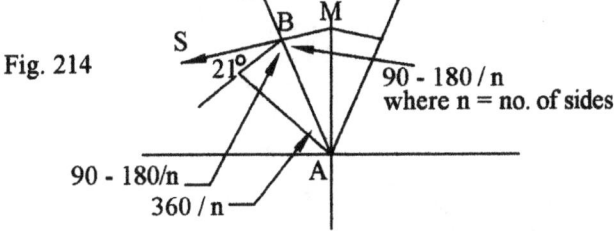

In Fig. 213, CA bisects ∠ DAT, so ∠ DAC = ∠CAT = 90 - a
    ∠ TAM is the supplement, so ∠ TAM = 2a.
Thus ∠ TAM = 2a = ∠VMS.
This makes RT ‖ VN, since they both form congruent corresponding angles.

21. Refer to Fig. 214.

Fig. 214

At point B, ∠ MBA + ∠ CBA = 180 - 21, or 90 - 180/n
+ 90 - 180/n = 159, which makes n = 0.44
Therefore, such a polygon cannot exist.

22.

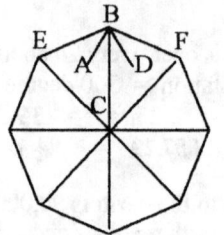

Fig. 215

$\angle$ ECB = 360 / 8 = 45 deg
$\angle$ EBC = $\angle$ BEC = (180 - 45) / 2 = 67.5 deg
$\angle$ ABD = $\angle$ ABC + $\angle$ DBC = 67.5 deg

Fig. 216

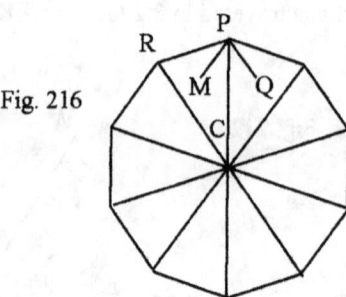

$\angle$ RCP = 360 / 10 = 36 deg
$\angle$ MPC = $\angle$ QPC = (180 - 36) / 2 = 72 deg
$\angle$ MPQ = 72 + 72 = 144 deg

24. Refer to Fig. 217
SM = SV, as each is 1/2 of the equal sides of an equilateral triangle.
$\angle$ 1 = $\angle$ 2 , since they are opposite the equal sides of $\triangle$ MSV.
$\triangle$ STU is equilateral, so $\angle$ S = 60 deg.

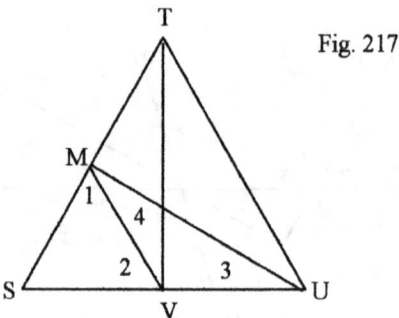

Fig. 217

thus $\angle$ 1 = $\angle$ 2 = 60 deg.
By SAS, $\triangle$ SMU $\cong$ $\triangle$ TMU, so $\angle$ SMU = 90 deg.
This makes $\angle$ 4 = 30 deg.

Fig. 218

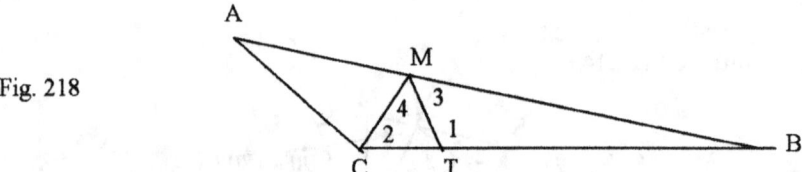

25. Refer to Fig. 218
$\angle$ A + $\angle$ B + $\angle$ C = 180
        where $\angle$ B = $\angle$ A / 2 and $\angle$ C = 3 $\angle$ A
$\angle$ A + $\angle$ A /2 + 3 $\angle$ A = 180. $\angle$ A = 40 deg.
$\angle$ B = 20, $\angle$ C = 120, and $\angle$ 2 = 60, since CM bisects $\angle$ C.
If $\angle$ B = 20, and $\angle$ 2 = 60, $\angle$ 3 + $\angle$ 4 = 100, and $\angle$ 3 = 50, since TM bisects $\angle$ CMB.
Thus $\angle$ B = 20 deg, $\angle$ 3 = 50 deg, and $\angle$ 1 = 110 deg.

26.

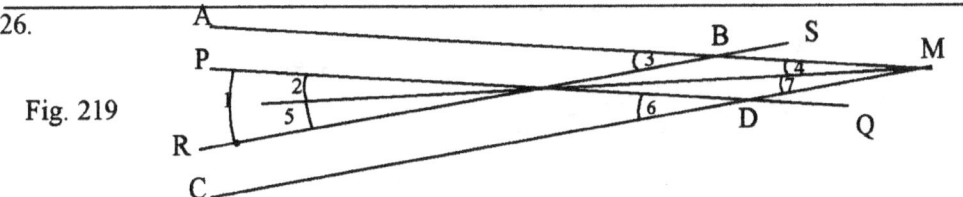

Fig. 219

1. Draw PQ || AB.   RS || CD (construction).
2. ∠2 = ∠3 = ∠4 (corr. ∠ s of || lines are ≅).
3. ∠5 = ∠6 = ∠7 (corr. ∠ s of || lines are ≅).
4. ∠2 = ∠4, ∠5 = ∠7 (trans. prop.)
5. ∠2 +∠5 = ∠4 +∠7 (addition).
6. ∠1 = ∠ ABC, the desired ∠ (subst.)
(Note that the problem does not allow the extension of AB and CD until they intersect. That is not allowed. Actually, the extenson of both AB and CD was not needed in this solution. The extension shown in this solution was done only to prove, show, and justify the correctness of the solution).

27.   Fig. 220

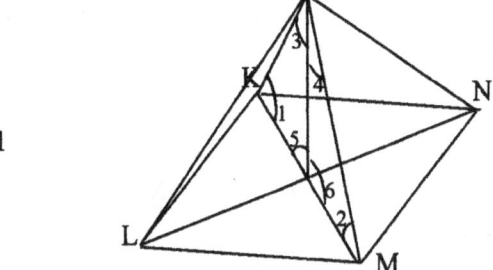

1. KLMN is a square and is basically a parallelogram, and the diagonals of a parallelogram bisect each other.
2. By SSS, △KYM ≅ △MYN.
3. ∠ 1≅∠ 2   (CPCTC or corr angles of congruent triangles are congruent).
4. NY ⊥ KM (If two congruent adjacent angles have their exterior sides on a line, the angles are right angles formed only by perpendicular lines).

28. Refer to Fig. 221

Fig. 221

1. The lateral faces are equilateral triangles (given).   2. ∠1 = ∠2 (base angles of isosceles triangle KVM are equal).   3. Draw VY, the bisector of ∠KVM (an angle can be bisected). 4. ∠3 = ∠4 (def. of bisector). 5. ∠5 = ∠6 (if 2 angles of a triangle are equal to two angles of another, the third angles are equal).   6. VY ⊥ KM (if the external sides of two congruent adjacent angles lie on a line, the angles are right angles).

29. P = 3x - 10 + 2x + 10 + 6x - 20 = 90;   x = 10, 3x - 10 = 20 ; 2x + 10 = 30 ; and 6x - 20 = 40. To find the area, use Hero's formula, namely:

$$A = \sqrt{s(s-a)(s-b)(s-c)}$$ where s = P/2 = 90/2 = 45

hence A = $\sqrt{45(25)(15)(5)}$ = 290.47 sq. units.

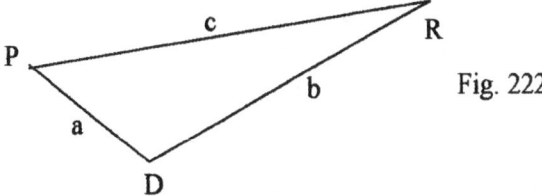

Fig. 222

30.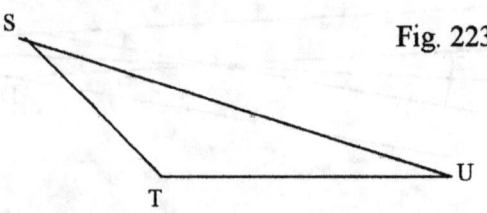
Fig. 223

Let side ST be x, TU as 1.5x, and US as 2x.   x + 1.5x + 2x = 126.   x = 28
1.5x = 42, and 2x = 56. Using Hero's formula:

$A = \sqrt{s\,(s-a)\,(s-b)\,(s-c)}$ where s = P /2 = 63.

$= \sqrt{63\,(35)\,(21)\,(7)}$ = 569.33 sq. units.

31.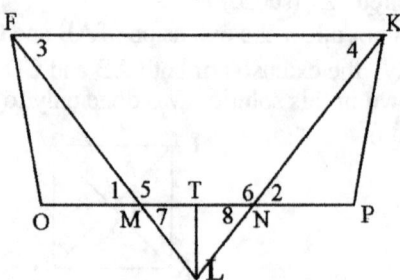
Fig. 224

q. FL = KL, ML = NL (sides of isosceles triangles FLK and MLN are equal)
2. $\angle 7 = \angle 8$ (base angles of an isosceles triangle are equal).   3. $\angle 7 = \angle 1$, $\angle 8 = \angle 2$
(vertical angles are equal)   4. $\angle 1 = \angle 2$ (trans. prop.)   5. OM = PN (given)
6. FL - ML = KL - NL (subt.)   7. FM = KN (subst.)   8. △FOM ≅ △KPN (SAS

32. Refer to Fig. 224.
1. ML = NL (sides of isosceles △MLN are ≅).   2. $\angle 7 = \angle 8$ (base angles of
an isosceles △ are =).   3. MT = NT (def. of median).
4. △LTN ≅ △LTM (SAS).

33.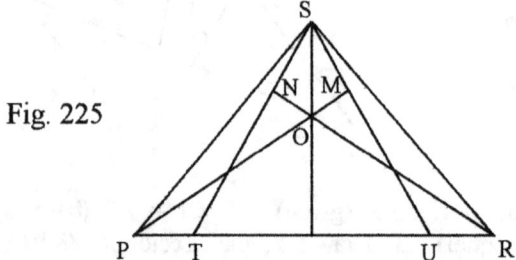
Fig. 225

1. In rt △s PMS and RNS, PT = RU, SM = SN, RN = PM (given).
2. △PMS ≅ △RNS (HL).   3. PS = RS, $\angle$SPT ≅ $\angle$SRU (CPCTC).
4. △PTS ≅ △RUS (SAS)

34.   Fig. 226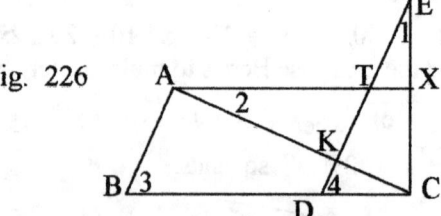

1. In rt △s BAC and DAC, AB = CD (given)
2. AB = DE (given)   3. $\angle 3$ ≅ $\angle 4$ (corr $\angle$ s of ‖ lines are ≅).
4. △ABC ≅ △DCE (HA).

35.

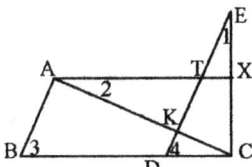

Fig. 226

1. AX ∥ BC, EC ⊥ BC, AC ⊥ AB, EC = AC, ED ∥ AB (given).   2. ∠3 ≅ ∠4 (corr. ∠s of ∥ lines are ≅).   3. △BAC ≅ △DCE (LA)   4. AB ≅ CD (CPCTC)

36. Refer to Fig. 226.
1. AB ∥ DE (given).   2. ∠3 ≅ ∠4 (corr. ∠s of ∥ lines are ≅).   3. ED ≅ AC, AC ⊥ AB, ED = BC (given).   4. △BAC ≅ △DCE (HA).

37.

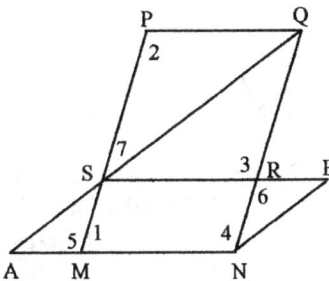

Fig. 227

1. PS = QR, PS ∥ QR (given).   2. PQRS is a ▱ (a pair of opp sides = and ∥).   3. △AMS ≅ △BRN (given).   4. SM ≅ NR, ∠A ≅ ∠B (CPCTC).   5. PS + SM = QR + NR (Add.)
6. PM ∥ RN (they lie on parallel lines).   7. PQNM is a ▱ (a pair of opp sides = and ∥).

8. ∠1 = ∠ PQN (opp ∠ s of a ▱ are ≅).   9. ∠ PQN = ∠PQA + ∠ARN (the whole is equal to the sum of its parts).   10. ∠ PQN ≅ ∠ 1 (alt int ∠s of ∥ lines are ≅)
11. ∠ PQA = ∠ B (trans. prop.).   12. ∠ B ≅ ∠ QSR (alt. int. ∠ s of ∥ lines are ≅).
13. ∠1 = ∠ B + ∠ SQR (subst.)

38.

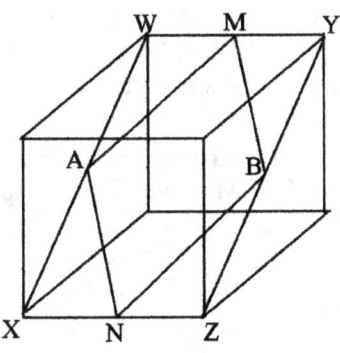

Fig. 228

1. WM = YM = XN = ZN (M, N, A, B are midpoints of congruent edges of the cube).
2. AM = BM = BN = AN (each is the hypotenuse of congruent right triangles, namely: △ AMN, △ BYM, △ NXA and △ NZB).   3. P_{MBNA} = AM + MB + BN + NA

$$= 2 \sqrt{(AM)^2 + (WM)^2} + 2 \sqrt{(VM)^2 + (YN)^2}$$
where AM = WX / 2, and WX = $e\sqrt{2}$ = 1.41 e, so AM = .71 e.
Hence, P_{MBNA} = 3.84 e.

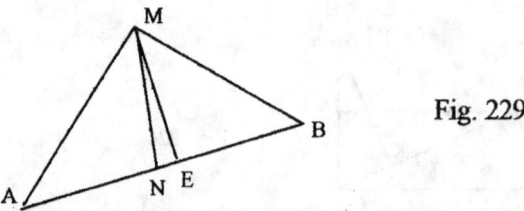

Fig. 229

1. △ MEN is a right triangle (ME is an altitude and therefore ⊥ to the base AB).
2. ∠ MEN = 90° (△MEN is a right △).   3. ∠MNE ≠ 90°, ⊥NME =   (a right triangle has one and only one right angle). 4. Hence, MN > ME (In any △, the side opposite the larger angle > the side opposite the smaller angle).

40.

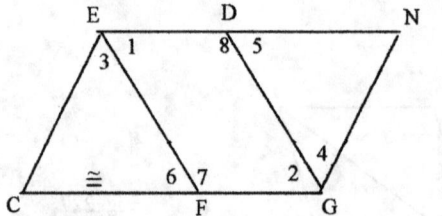

Fig. 230

1. DEFG is a ▱ (given).   2. ∠1 ≅ ∠ 2 (opp. ∠ s of a △ are ≅). 3. ∠1 = ∠3,
 ∠2 = ∠ 4 (def. of bisector).   4.   2 ∠ 1 = 2 ∠ 2 (mult). 4a. ∠ CEN = ∠CGN(subst.)
5. ∠1 ≅ ∠ 6 (alt int ∠s of ‖ lines are ≅).   6. ∠3 = ∠ 6 (trans. prop.) 7. ∠6 ≅ ∠2
(corr. ∠s of ‖ lines are ≅).   8. ∠2 ≅ ∠5 (alt. int. ∠ s of ‖ lines are ≅).
9. ∠ 6 = ∠ 5 (trans. prop.) 10. ∠ 5 = ∠ 4 (trans. prop.)   11. ∠C = ∠N (if 2 angles of one triangle are = to 2 angles of another triangles, the 3rd angles are equal).

41.

Fig. 231

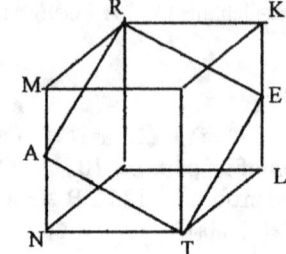

1. MA = AN = EL = KE (def. of midpoint. A and E are midpoints of equal edges
MN and KL of the cube).   2. MR = NT = LT = KR (all edges of the cube are equal).
3. △KMA ≅ △ TRA ≅ △ELT ≅ △EKR (LL).   RA = TA = ET = ER
(CPCTC)   5. RETA is a rhombus (all 4 sides are ≅).
42.
1. CW = HY = YI = WD (W, V, Y, and M are midpoints of equal sides).
2. VC = RY = BY = BJ (V, R, Y, B, M, J, and W are midpoints of equal sides.
3. △ VCW ≅ △ RHY ≅ △ BIY ≅ △ BTJ (LL).
4. VW = RY = WM = BY (CPCTC). 5. KV = KR = XM = XB (def. of midpoint) 6. △VKW ≅ △KRY ≅ △XBY ≅△XMW (LL).   7. KW = KY = XY
 = XW (CPCTC).   8. WKYX is a ▱ (all 4 sides are =).

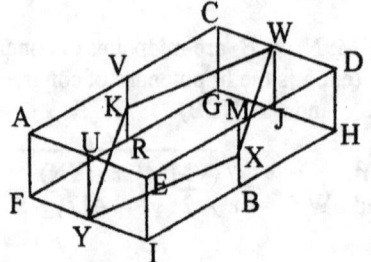

Fig. 232

43.

NOTE: The right end load tends to turn the beam counterclockwise about the fulcrum (point of support of the beam, see Fig. 233). The left end load tends to rotate the beam about the fulcrum clockwise.

The turning effect on the beam produced by these tendencies is called moment of force, and is equal to the product of force and the distance of its line of action from the fulcum. A counterclockwise moment is taken as positive, and the clockwise moment negative. For the beam to balance horizontally (or to be in equilibrium), the counterclockwise moments must equal the clockwise moments.

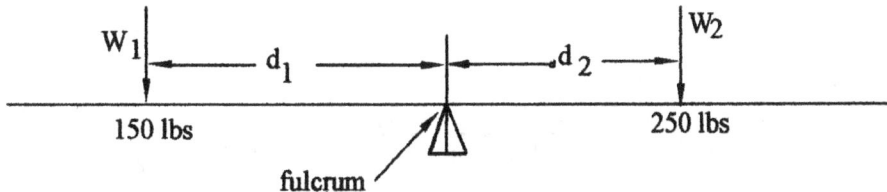

Counterclockwise moment: $W_1 \times d_1$ Clockwise moment: $W_2 \times d_2$

Applying the above concept, the counterclockwise moment must be equal to the clockwise moment. That is, $150 \times d_1 = 250 \times d_2$

The desired ratio is $d_1 / d_2 = 250 / 150 = 1.67$

44. Refer to Fig. 244. In the C scale, the freezing point of water is 0, and the boiling point is 100; in the F scale, the freezing point of water is 32 and the boiling point 212.

Note that the F scake reading starts at 32, not at 0. Therefore, 32 must
In the C scale, there are 100 (100 - 0) divisions or degrees, from
freezing to boiling points.

Correspondingly, there are 180 (212 = 32) divisions or degrees, from freezing to boiling points in the F scale.

We write these ratios as $C / F = 100 / 180$,

which yields the relationship $C = 5/9 \ F$.

be taken away from F, and the correct equation should be $C = 5/9 \ (F - 32)$.
If the equation is solved for F, $F = 9/5 \ (C) + 32$.

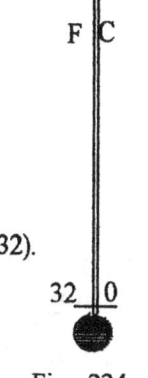

Fig. 234

45.

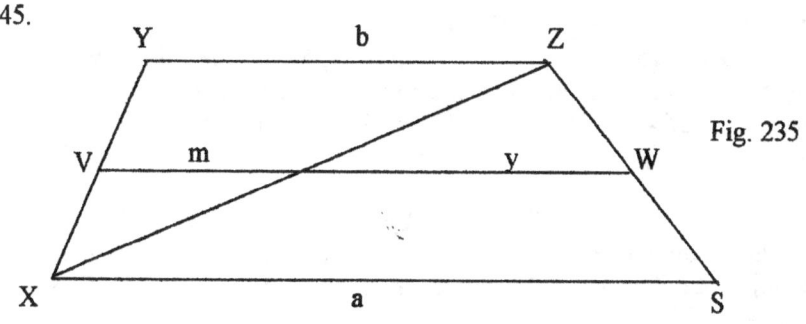

Fig. 235

1. In $\triangle YZX$, $m = b / 2$ (A line through the midpoint of the sides of a triangle is parallel to the third side and equal to 1/2 the length of that side.   2. In $\triangle XZS$, $y = a/2$ (same reason as in 1). Hence $m + y = b / 2 + a / 2 = 1/2 \ (a + b)$.

46.

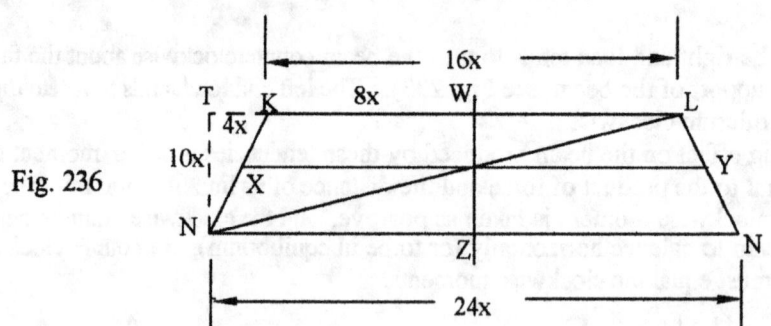

Fig. 236

1. KLMN is an isosceles trapezoid (given)   2. KW = 2x. MZ = 12x. then KT = 4x (by symmetry). 3. In triangle NTL, $(NL)^2 = (TL)^2 + (TN)^2$. (Use the Pythagorean Theorem). 4. Hence NL $= \sqrt{(20x)^2 + (10x)^2} = 22.36x$

47. Diagonals AM and BN of the cube are shown in detail in Fig. 237A for clarity. '
1. $\angle 1 \cong \angle 2$ (alt. int. $\angle$ s of ‖ lines are $\cong$).   2. $\angle 3 \cong \angle 4$ (vert. $\angle$ s are $\cong$). $\triangle$AVB $\sim \triangle$MVN (AA).   4. AB / MN = AV / MV (corr. sides of $\sim \triangle$s are in prop. or CSSTP). 5. AB (MV) = MN (AV)   (Law of proportion).

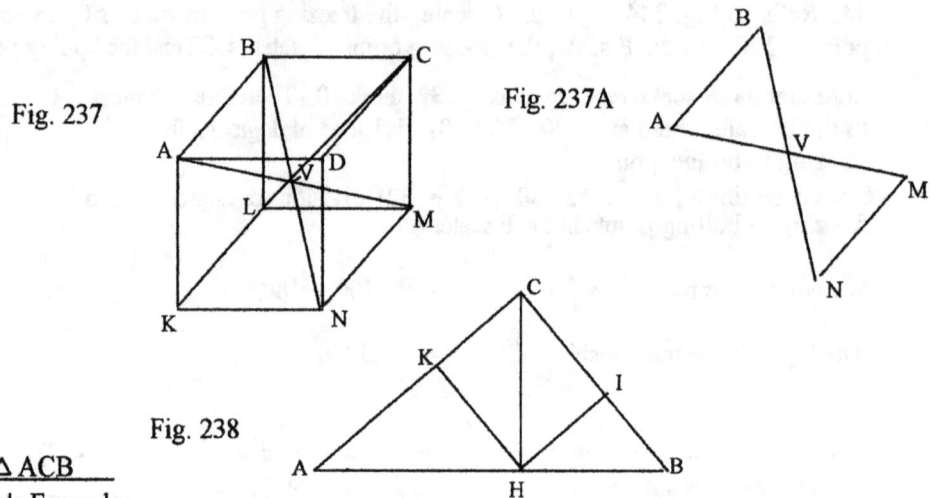

Fig. 237

Fig. 237A

Fig. 238

For rt $\triangle$ACB
By Hero's Formula:
$A = \sqrt{s(s-a)(s-b)(s-c)}$,   where s = (a + b + c) / 2   =   30
$= \sqrt{30(5)(10)(15)}$ =   150

Also, A = (1/2) bh; 150 = (1/2) 25 CH ; CH = 12
1. Let AH = x, and HB = 25 - x; then x / AH = AH / (25 - x) (the altitude to the hypotenuse of a right triangle is the geometric mean between the segments of the hypotenuse). Solving the proportion, x = 16 and 25 - x = 9.   Thus, AH = 16 and HB = 9.

For rt $\triangle$ AHC
2. Let CK = x, and KA = 20 - x; then CA / CH = CH / CK, or 20 / 12 = 12 / x (either leg of a right triangle is the geometric mean between the hypotenuse and the segment of the hypotenuse adjacent to that leg). Solving the proportion, x = 7.2 and 20 - x = 12.8. Thus CK = 7.2 and KA = 12.8.

3. Let KH = x; then CK / KH = KH / KA, or 7.2 / x = x / 12.8. Solving the proportion, x = 9.6. Thus, KH = 9.6.

For rt $\triangle$ BHC:

48.

4. Let BI = x. amd IC = 15 - x; then BC / BH = BH / x   or   15 / 9 = 9/ x   (either leg of a right triangle is the geometric beam betweeen the hypotenuse and the segment of the hypotenuse adjacent to that leg). Solving the proportion, x = 5.4,   15 - x = 9.6.
geometric mean between the segments of the hypotenuse).   Solving the proportion, x = 7.2. Thus
5. Let HI = x; BI / HI = HI / IC, or 5.4 / x = x / 9.6   (the altitude to the hypotenuse is the
HI = 7.2.

49.

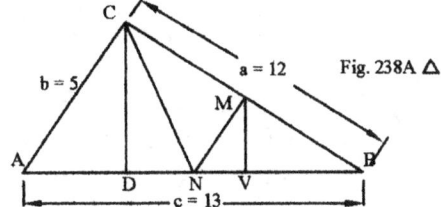

Fig. 238A $\triangle$

1. In rt $\triangle$ACB, MN = 1/2 (AC) (A line that passes through the midpoint of two sides of a triangle is parallel to the third side and equal to 1/2 of that side). Thus MN = (1/2) 5 = 2.5.
2. By Hero's Formula, the area of rt $\triangle$ACB   A = $\sqrt{s(s - a)(s - b)(s - c)}$, where s = (a + b + c) / 2 = 15.   A = $\sqrt{15\ (3)\ (10)\ (2)}$ = 30
Thus A = (1/2) b h or 30 = (1/2) 13 (CD)   and   CD = 4.62.
3. In rt $\triangle$CDH, MV = (1/2) CD = 2.31 (same reason as in 1).
4. $(MV)^2$ + $(NV)^2$ = $(NM)^2$;   MV = .95.

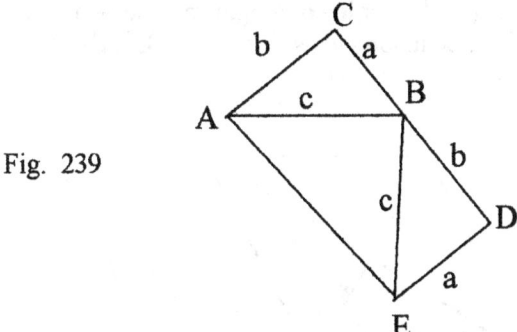

Fig. 239

50. President Garfield's proof of the Pythagorean Theorem:
1. Area of the trapezoid ACDE = (1/2) (a + b) (a + b) (Aea of a trapezoid = (1/2 of the sum of the bases x height).   2. From the figure, the Area of the trapezoid = areas of the two congruent right triangles ACB and EDB + the area of the isosceles right triangle ABE.

3. (1 / 2) (a + b) (a + b) = 2 x (1 / 2) ab + (1 / 2) c^2
Solving the equation yields c^2 = a^2 + b^2.

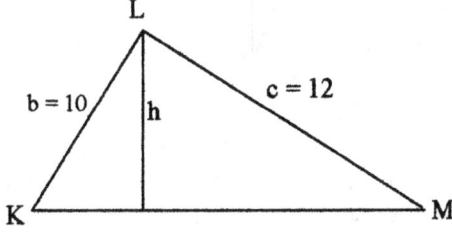

Fig. 240

51. By Hero's Formula,   A = $\sqrt{s(s - a)(s - b)(s - c)}$ where s = (1/2) (a + b + c) = 20
= $\sqrt{20\ (2)\ (10)\ (8)}$ = 56.57

A = (1/2) b h = (1/2) 18 (LT) ,   h = LT = 6.26

52.

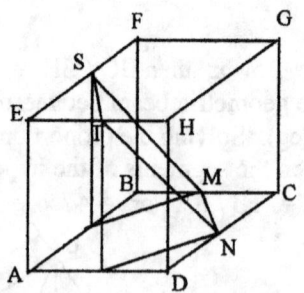

Fig. 241

$ST = MN = 5\sqrt{2}$ $UM = VN = 5\sqrt{2}$
$US = VT = 10$
$SM = TN = \sqrt{(10)^2 + (5\sqrt{2})^2}$
$= 12.25$

P of TSMN = ST + MN + SM + TN = 7.05 + 7.05 + 12.25 + 12.25 = 38.63.

53. Refer to Fig. 241 $ST = MN = 9\sqrt{2}$ $UM = VN = 9\sqrt{2}$
$US = VT = 18$ $SM = TN = \sqrt{(18)^2 + (9\sqrt{2})^2} = 22.05$
P of TSMN = ST + MN + SM + TN = 12/69 + 12.69 + 22.05 + 22.05 = 69.48

54. Refer to Fig. 242.

In rt $\triangle$SMV, $SM = x\sqrt{3}$ (the longer leg of a right triangle = the shorter leg x $\sqrt{3}$).
Thus 9 = x   3. Solving the equation yields x = 3   3, 2x = $6\sqrt{3}$.
The desired P is 4 (2x) = 4 $(6\sqrt{3})$ = 41.52.

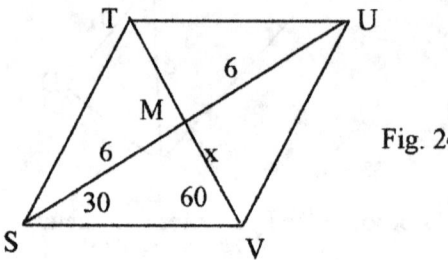

Fig. 242

55. The Pythagorean Theorem derived.

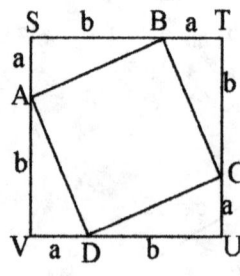

Fig. 243

In Fig. 243, note that the area of the big square = the sum of the areas of the small square
plus the area of the 4 congruent triangles.
A of STUV = $(a+b)(a+b)$ = $a^2 + 2ab + b^2$
A of the 4 congruent triangles = 4 x (1/2) ab = 2 ab
A of the small square ABCD = c^2

Thus, $a^2 + 2ab + b^2$ = 2 ab + c^2
Finally, subt. 2ab from both sides of the equation yields $a^2 + b^2 = c^2$.

56.

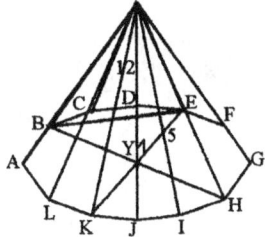

Fig. 244

BYE is a 45-45 right triangle, so BE = YE $2\sqrt{ } = 5$ (1.41) = 7.05/ The lateral edge BV =
EV = $\sqrt{(BY)^2 + (VY)^2}$ = 13.
By Hero's Formula, $A = \sqrt{s\,(s-a)\,(s-b)\,(s\sqrt{c})}$ where s = (a + b + c) / 2 = 16.53.
= $\sqrt{16.53\,(11.53)\,(11.53)\,(9.48)}$ = 144.33.

57.

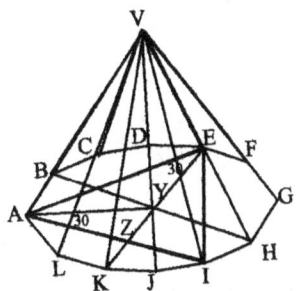

Fig. 245

AEI is an equilateral triangle.   Area of an equilateral triangle is $(s^2/4)\sqrt{3}$, where s is the length
of a side.   AZY is a 30-60 right triangle, so AY = 5, ZY = 2.5, and AZ = 2.5 $\sqrt{3}$. Thus
AI = 5 $\sqrt{3}$ = 8.75.   Hence, the area of triangle AEI = $(s^2/4)\sqrt{3}$ = $[(8.75)^2/4]\sqrt{3}$
= 33.11.

58.

Fig. 246.

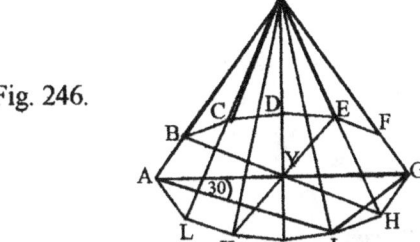

AIG is a 30-60 right triangle.   AY = 5 (given), so Y/Z = 2.5 and AZ = 2.5 $\sqrt{3}$.
AI is twice AZ, thus AI = 5 $\sqrt{3}$ = 8.65.
AG = 10 (given), so IG = 5. Hence the area of triangle AIG = (1/2) b h
= (1/2) 8.65 (5) = 21.63.

59.

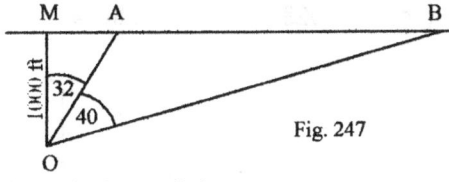

Fig. 247

Speed of train = dist AB / travel time
AB = MB - MA   MB /1000 = tan 72 ;   MB = 1000 tan 72 = 3080
MA / 1000 = tan 32;   MA = 1000 tan 32 = 620
AB = 3000 - 620 = 2460.   Speed of train = 2460 / 10 = 246 ft /sec.

60. Refer to Fig. 247, prob. 59
Speed of train = dist AB / travel time
            = 2460 / 15 = 164 ft / sec.

61.

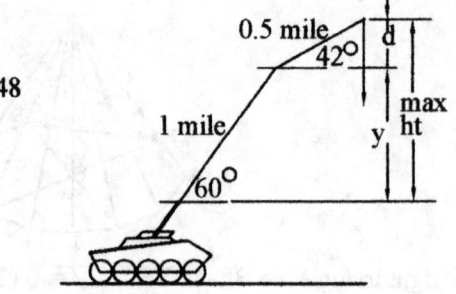

Fig. 248

Max ht = y + d (see Fig. 248)
y / 1 = sin 60   y = 1 (.87) =   .87 mi.
d / .5 = sin 42   d = ,5 sin 42 =   .34 mi.
max ht = y + d   =   1.21 mi.

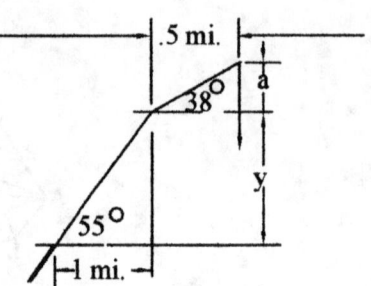

62.

Fig. 249

Max alt. = y + a
y / 1 = tan 55   y = 1.43 mi.
a / .5 = tan 38   a = .39 mi.
Max alt = 1.43 + .39 = 1.82 mi.

63. Refer to Fig. 249 prob. 62
Part 1 of flight

Dist = hor speed x time
○   where hor speed = 15000 cos 55   =   8550 ft / sec
5280 = 8550 (t) ;   t = .62 sec.

Part 2 of flight

Dist = hor speed x time
        where hor speed =   15000 cos 38
                        =   11850 ft / sec
2640 = 11850 (t);   t = .22 sec.

Thus, time for the rest of the flight of the missile = [8 - (.62 + .22)] = 7.16 sec.
Hence, dist trav in 7.16 sec =   11850 (7.16) =   84846 ft
Total dist trav at the end of 8 sec =   5280 + 2640 + 84846 =   92766 ft.

64.

Fig. 250

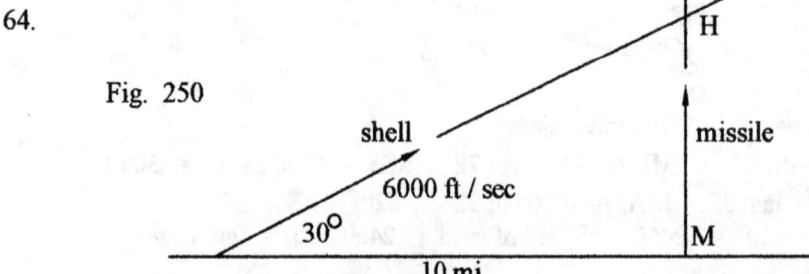

Velocity of shell, horizontal = 6000 cos 30
                                =   5220 ft / sec.

$52800 = 5220 \text{ (t)}, \quad t = 10.11 \text{ sec}$

From the figure, $MH / 10 = \tan 30$
$MH = 52800 \, (.58) = 30624 \text{ ft}$

Dist = speed x time
$30624 = \text{speed } (10.11)$; hence, the required speed = $30824 / 10.11 = 3029.8 \text{ ft / sec}$.

65. The direct hit will occur at 30,624 ft altitude. (See the height MH in prob. 64).

66. Altitude = vertical speed x time = $800 \, (10.11) = 8088 \text{ ft}$
This means that the shell will not hit the missile after all. At the instant that the hit is expected to occur, the shell is 30,624 ft high while the missile is only at an altitude of 80088 ft.

67. From prob. 64, the vertical speed of the shell is 3029.08 ft / sec, and the direct hit on the missile would have occurred at an altitude of 30,624 ft. The hit would happen 10.11 seconds after the instant of firing. (Altitude - vertical speed x time; $30624 = 3029.08$ x t, $t = 10.11$ sec) so, if the missile is fired 1 / 2 sec after the firing of the shell, there will be no direct hit, since the missile will be at an altitude of 29,109.46 ft as shown below:

Altitude = speed x time = $3029.08 \, (10.11 - 0.5) = 29,109.46 \text{ ft}$

68.

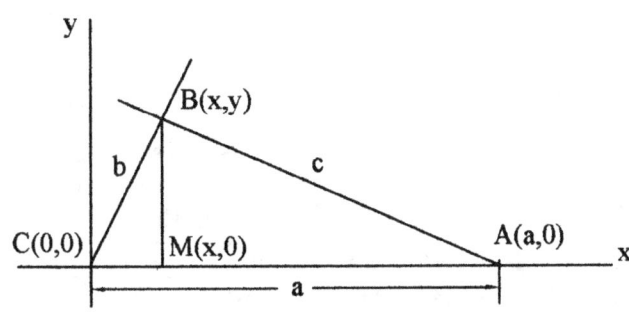

In $\triangle$ BMA:
$$\begin{aligned} c^2 &= (a - x)^2 + h^2 \\ &= a^2 - 2ax + x^2 + (b^2 - x^2) \\ &= a^2 - 2ab \cos C + b^2 \text{, or} \\ c^2 &= a^2 + b^2 - 2ab \cos C \text{ (the Law of Cosines).} \end{aligned}$$

69.

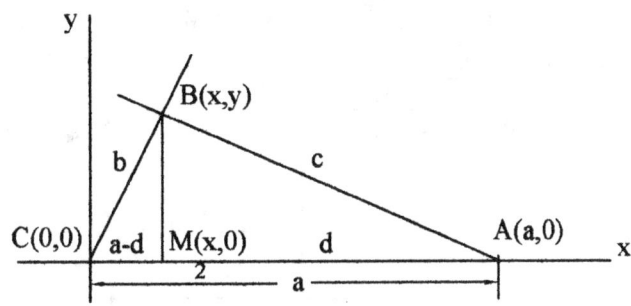

From the Law of Cosines, $c^2 = a^2 + b^2 - 2ab \cos C$
From the figure,
$$\begin{aligned} c^2 &= a^2 + b^2 - 2ab \, [(a - d) / b] \\ &= a^2 + b^2 - 2a^2 + 2ad \\ &= b^2 - a^2 + 2ad. \end{aligned}$$

70.

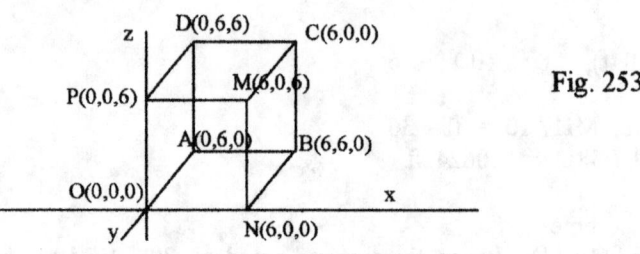

Fig. 253

From the figure, $l = 6$, $w = 6$, $h = 6$; V of solid $= l \times w \times h = 6 = 216$ cu. in. $= .13$ cu. ft. V of water displaced $=$ V of solid $= .13$ cu. ft. [3]

71.

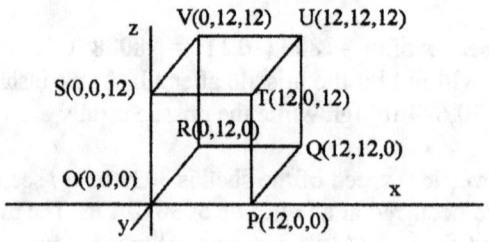

Fig. 254

V of solid $= 216$ cu. in. $=$ V of water displaced.
Area of water surface $= 12 \times 12 = 144$ sq. in.
When the solid is immersed into the water, 216 cu. in. of water will be displaced, and the water level rises to a height h.
To find h: $V = 144 \times h$, and $h = 216 / 144 = 1.5$ in.

72.

Fig. 255

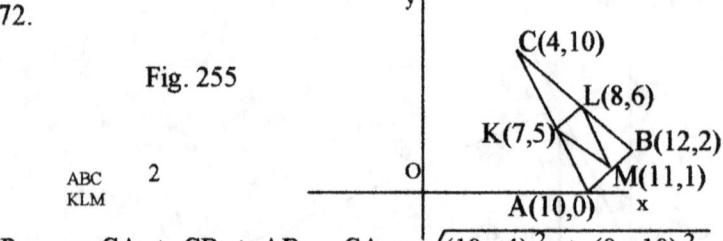

ABC   2
KLM

$P_{ABC} = CA + CB + AB$; $CA = \sqrt{(10 - 4)^2 + (0 - 10)^2} = 11.6$.
$CB = \sqrt{(12 - 4)^2 + (2 - 10)^2} = 11.31$ $AB = \sqrt{(12 - 10)^2 + (2 - 0)^2} = 2.83'$
$P_{ABC} = 25.8$.

$A_{ABC} = \sqrt{s\,(s - a)\,(s - b)\,(s - c)}$ where $s = (1/2)\,(a + b + c) = 12.9$
$= \sqrt{12.9\,(1.24)\,(1.59)\,(10.07)} = 16$.

As in P above, using the dist formula:
$KL = \sqrt{2} = 1.41$, $KM = \sqrt{32} = 5.66$, $LM = \sqrt{34} = 5.83$. $P_{KLM} = 12.9$.

As in A above, using Hero's Formula: $s = (1/2)\,P = (1/2)\,12.9 = 6.45$
$A_{ABC} = \sqrt{6.45\,(.79)\,5.04\,(.62)} = 3.99$.

Hence, $P_{ABC} / P_{KLM} = 25.8 / 12.9 = 2$; $A_{ABC} / A_{KLM} = 16 / 3.99 = 4.01$

73.

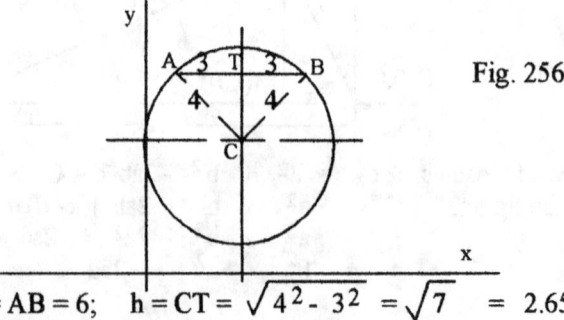

Fig. 256

$A_{ABC} = (1/2)\,bh$; $b = AB = 6$; $h = CT = \sqrt{4^2 - 3^2} = \sqrt{7} = 2.65$.
hence $A = (1/2)\,6\,(2.65) = 7.95$ sq. in.

74.

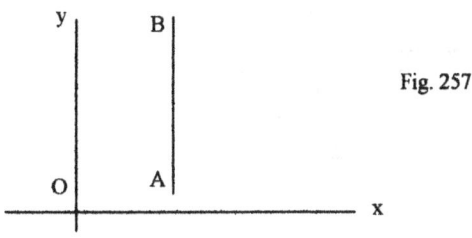

Fig. 257

The slope of AB is undefined. Here's why:

m = slope = $\Delta y / \Delta x$, where $\Delta x = 0$ since the line is vertical. Assume $\Delta x = 1/16$, $1/32$, $1/64$ ... diminishing swiftly, approaching zero, while the value of $\Delta y / \Delta x$ soars infinitely until it becomes immeasurable, so infinitely large that there is no end to its increase ... it has no upper limit, it cannot be defined.

75.

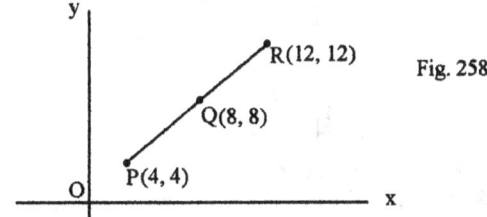

Fig. 258

Another method, instead of using the slopes of both segments, would be to show that PQ + QR = PR, since the equation / statement can only be true if all three points P, Q, and R are collinear. If they are not in line, obviously the sum of PQ and QR would be longer than PR.

$PQ = \sqrt{(8-4)^2 + (8-4)^2} = 4\sqrt{2}$
$QR = \sqrt{(12-8)^2 + (12-8)^2} = 4\sqrt{2}$
$PR = \sqrt{(12-4)^2 + (12-4)^2} = 8\sqrt{2}$

The above calculations show that PR = PQ + QR. Therefore, the points P, Q, and R all lie on the same line.

76. All points P(x, y) that are equidistant from S and T must lie on the perpendicular b--sector of ST.

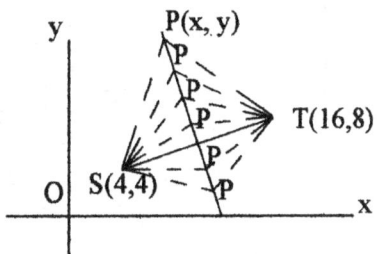

That is, for all P, PS = PT.

$$\sqrt{(x-4)^2 + (y-4)^2} = \sqrt{(x-16)^2 + (y-8)^2}$$

Squaring both sides and combining terms yields the desired equation: $3x + y = 36$. (This can be checked by using the point-slope form of the equation of a line).

77.

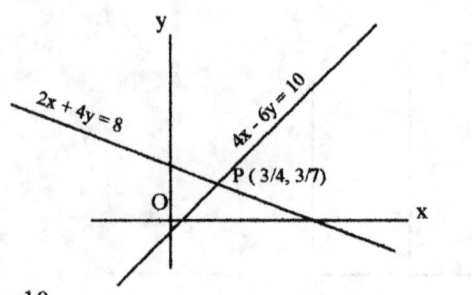

Fig. 260

The lines   4x - 6y = 10
                2x + 4y = 8,   when solved simultaneously yield intersection point
P (3/4, 3/7).   Hence the desired equation of a line parallel to the x-axis and passing
through P is: y = 3 / 7

78.

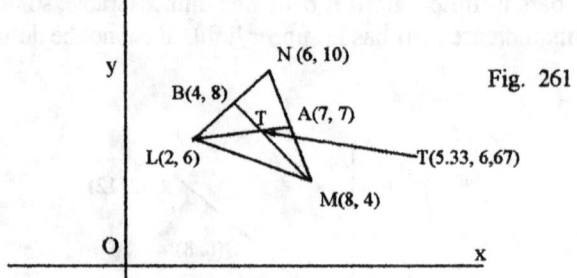

Fig. 261

m_{LA} = (7 - 6) / (7 - 2) = 1 / 5 ;   Equation of LA: y - 7 = 1 / 5 (x - 7)   or   x - 5y = - 28
m_{MB} = (8 - 4) / (4 - 8) = - 1 ;   Equation of MB: y - 4 = -1   (x - 8)   or   x + y = 12
Solving both equations simultaneously,   x = 5.33,   y = 6.67

Hence, the equation of the line through T and parallel to MN is : m_{MN}= (10 - 4) / (6 - 8)
= -3.   y - 6.67 = -3 (x - 5.33)   or   y + 3x = 22.66.

79.

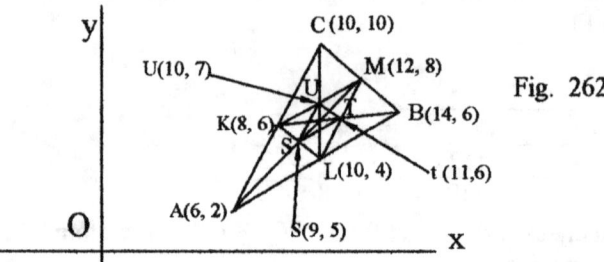

Fig. 262

S, T, and U are the midpoints, respectively, of KL, LM, and MK. Hence, these midpoints are S(9, 5), T(11, 6) and U(10, 7).

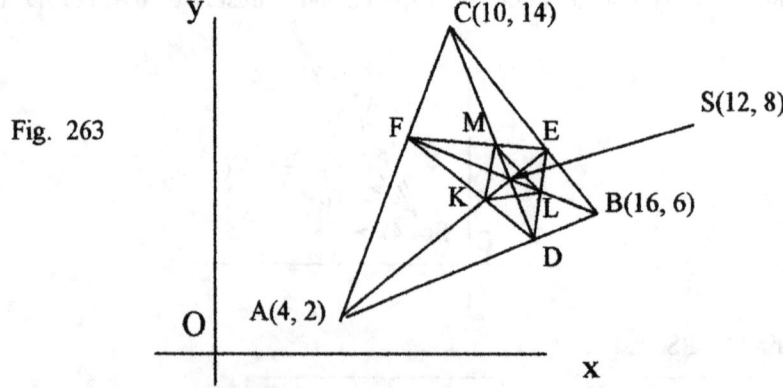

Fig. 263

m_{BC} = (14 - 6) / (10 -16) = 8 / -6 = -4 /3 ; so m_{AE} = 3 / 4
Eq. of alt. AE: y - 2 = (3 / 4) (x - 4)
M_{AB} = (6 - 2) / (16 - 4) = 1 / 3 ; m_{CD} = -3; Eq. of alt. CD: y - 14 = -3 (x - 10)
Solving the two equations simultaneously: x = 12, y = 8. The orthocenter is at S(12, 8).

80a.

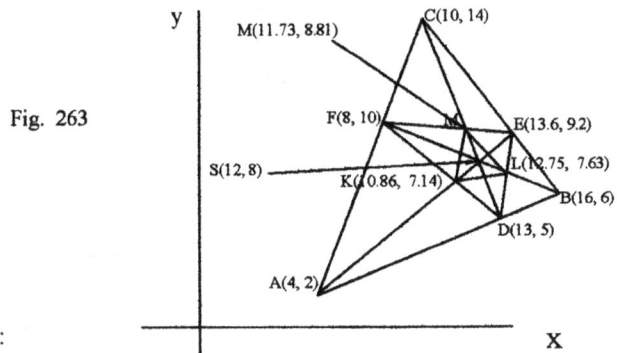

Fig. 263

Coordinates of F:

$m_{AC} = (14 - 2) / (10 - 4) = 2.$ $m_{BF} = -1/2$
Eq. of BF: $y - 6 = -1/2 (x - 16)$ (1)
Eq. of AC: $y - 14 = 2(x - 10)$ (2)
Solving eq. (1) and (2) simultaneously, x = 8, y = 10, thus F is at (8, 10)

Coordinates of E:

$m_{BC} = (6 - 14) / (16 - 10) = -4/3;$ $m_{AE} = 3/4$
Eq. of BC: $y - 6 = -4/3 (x - 16)$ (3)
Eq. of AE: $y - 2 = 3/4 (x - 4)$ (4)
Solving eq. (3) and (4) simultaneously, x = 13.6, y = 9.2

Coordinates of D:

$m_{AB} = (6 - 2) / (16 - 4) = 1/3;$ $m_{CD} = -3$
Eq. of AB: $y - 6 = 1/3 (x - 16)$ (5)
Eq. of CD: $y - 14 = -3 (x - 10)$ (6)
Solving eq. (5) and (6) simultaneously, x = 13, y = 5

Coordinates of M:

Eq. of FE: $y - 10 = [(10 - 8.5) / (8 - 12.67)] (x - 8)$
Eq. of CD: $3x + y = 44$
Solving the two equations simultaneously:
x = 11.73, y = 8.81,   thus M is at (11.73, 8.81).

Coordinates of L:

Eq. of BF: $x + 2y = 28$
Eq. of DE: $y - 5 = [(8.5 - 5) / (12.67 - 13)] (x - 13)$
Solving both equations simultaneously:
x = 12.75, y = 7.63.

Coordinates of K:

Eq. of FD: $y - 10 = [(10 - 5) / (8 - 13)] (x - 8)$
Eq. of AE: $3x - 4y = 4$
Solving the two equations simultaneously:
x = 10.86, y = 7.14

P of KLM:

$KM = \sqrt{(11.73 - 10.86)^2 + (8.81 - 7.14)^2} = 1.88$

$ML = \sqrt{(12.75 - 11.73)^2 + (7.63 - 8.81)^2} = 1.57$

$KL = \sqrt{(12.75 - 10.86)^2 + (7.63 - 7.14)^2} = 1.95$

Hence, P of triangle KLM = KM + ML + KL = 5.4

80b. Equation of the line through orthocenter S parallel to AB: (Refer to Fig. 263).
$m_{AB} = (6 - 2) / (16 - 4) = 1/3$, hence the desired equation is: $y - 8 = 1/3 (x - 12)$
or x - 3y = -12.

81.

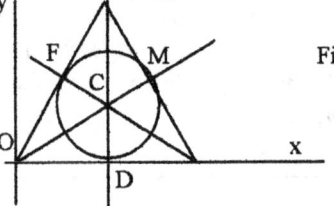

Fig. 264

81.

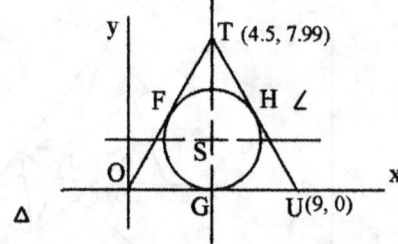

Fig. 264

1. OUT is an equilateral triangle with altitudes TG, FU, and HO. The altitudes bisect the three angles of the triangle, so $\angle 1 = \angle 2 = \angle 3 = 30$ deg, and $\triangle$TGO, $\triangle$OFU, and $\triangle$UFO are 30-60 right triangles. 2. In right triangle TGO, TG = OG $\sqrt{3}$ = 4.5 $\sqrt{3}$ = 7.79, so T has coordinates (4.5, 7.79). 3. Eq. of line TO: y - 0 = (7.79) / 4.5) (x - 0), or y = 1.73x.

4. Eq. of UF: y - 0 = (-1 / 1.73) (x - 9) or y = -.58x + 5.22. Solving both equations simultaneously, x = 2.26, y = 3.91. F is at (2.26, 3.91). 5. Eq. of UT: y - 0 = (7.79 - 0) / (4.5 - 9) (x - 3) or y = -1.73x + 15.59.
Eq. of OH: y - 0 = -(-1 / 1.73) (x - 0) or y = .58x. Solving both equations simultaneously x = 6.74, y = 3.91. H is at (6.74, 3.91).

82,

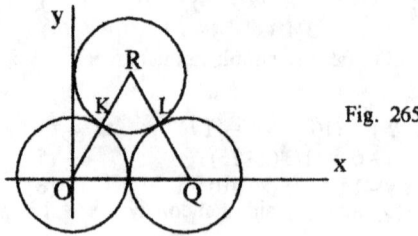

Fig. 265

Solution 1: By symmetry, the intersection points K and L must lie on the midpoints of the sides OR and QR of the triangle. Hence, the tangency point K (x_m, y_m) would be at: *x_m = (0 + 6) / 2 = 3; *y_m = (0 + 6 $\sqrt{3}$) / 2 = 3 $\sqrt{3}$.

Solution 2: To locate K and L, find the intersections of circle O and side OR, and the circle Q and side QR.
a. Eq. of OR: y - 0 = 3 (x - 0) or y = 3x.
Eq. of circle: $(x - 0)^2 + (y - 0)^2 = 36$
Solving both equations simultaneously, x = 3, y = 3 $\sqrt{3}$. Thus K is at (3, 3 $\sqrt{3}$)
b. Eq. of QR: y = - $\sqrt{3}$ (x - 12) or y = - $\sqrt{3}$x + 12 $\sqrt{3}$.
Eq. of circle: $(x - 12)^2 + (y - 0)^2 = 36$.
Solving both equations simultaneously: x = 9, y = 3 $\sqrt{3}$. Thus L is at (9, 3 $\sqrt{3}$).

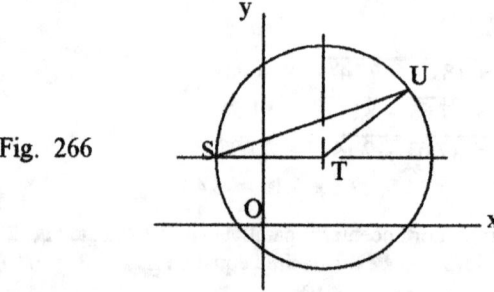

Fig. 266

To locate U: Eq. of SU: By the distance formula, $(6 \sqrt{10})^2 = (x + 4)^2 + (y - 6)^2$ or $x^2 + 8x + y^2 - 12y = 308$. Eq. of UT: $10^2 = (x - 6)^2 + (y - 6)^2$, or $x^2 - 12x + y^2 - 12y = 28$. Solving both equations simultaneously: x = 14, y = 12. Hence, U is located at (14, 12).

---

* midpoint

84.

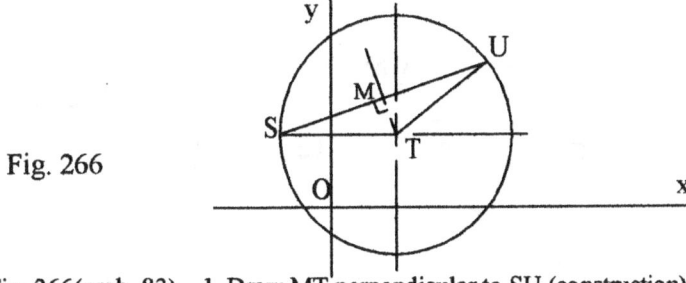

Fig. 266

Refer to Fig. 266(prob. 83). 1. Draw MT perpendicular to SU (construction).
2. In the right triangle TMU, $(MT)^2 + (MU)^2 = 10^2$; MT = 3.6
Hence, A $= (1/2)$ SU (MT) $= (1/2)$ 6 $\sqrt{10}$ (3.16) $= 30$ sq. in.

85. (Fig. 257). If OS = OP, the circle would pass through both S and P.
But: OP = $\sqrt{(15-0)^2 + (12-0)^2} = 19.21$, and
OS = $\sqrt{(17-0)^2 + (14-0)^2} = 22.02$. Thus, OS > OP. The answer is NO.

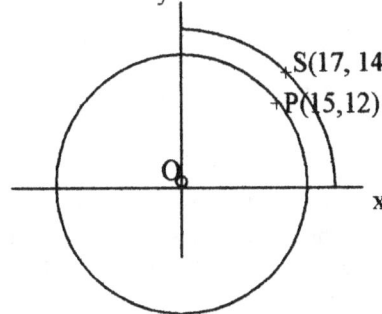

Fig. 257

86.

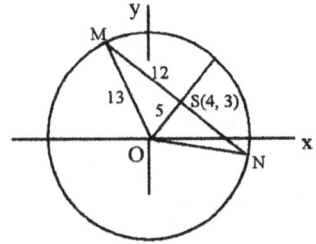

Fig. 268

The coordinates of M and N are really the intersections of MN and OM, and of MN
and ON. OS = $\sqrt{(4-0)^2 + (3-0)^2} = 5$. In the right triangle MSO,
OM = 13 by the Pythagorean Theorem. Eq. of OM: $13^2 = (x-0)^2 + (y-0)^2$
or $x^2 + y^2 = 169$.
$m_{OS} = 3/4$; $m_{MN} = -4/3$.
Eq. of MN: $y - 3 = -4/3 (x-4)$ or $y = (25-4x)/3 = 8.33 - 1.33 x$
Solving the equations of OM and MN simultaneously:
$169 - x = (8.33 - 1.33x)$, and by the quadratic formula, x = -3.21. So y = 12.60.
Thus the coordinates of M are: (-3.21, 12.60).

Eq. of ON: $13^2 = (x-0)^2 + (y-0)^2$ or $x^2 + y^2 = 169$.
Eq. of MN: $y = -1.33x + 8.33$. Solving both equations simultaneously: x = 11.21,
and y = -6.58.

87. Equation of the circle in prob. 86: Note that the equation of any circle with center
at (h, k) and radius r is $(x-h)^2 + (y-k)^2 = r^2$. Thus the desired equation of the
circle in the preceding problem is: $(x-0)^2 + (y-0)^2 = 13^2$.

88. Refer to Fig. 268, prob. 86. Eq. of OSK: $y - 0 = 3/4 (x-0)$ or $y = (3/4)$ x.
Eq. of circle O: $x^2 + y^2 = 13^2$. Solving both equations simultaneously, x = 10.4,
y = 7.8.

89.

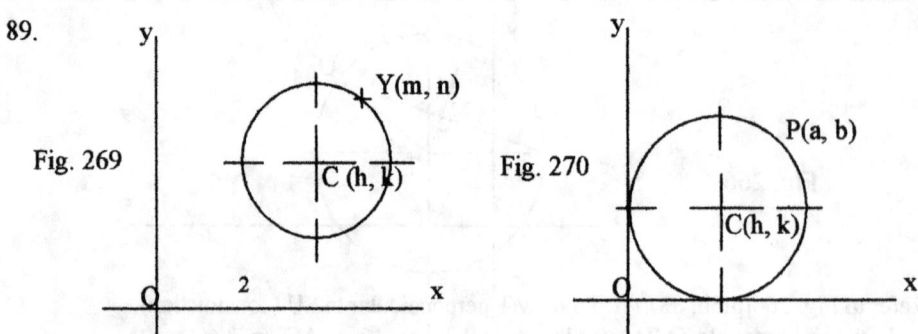

Fig. 269

Fig. 270

In conjunction with solution No. 87, the equation of any circle is: $(x - h)^2 + (y - k)$ $= r^2$, where (h, k) is the center of the circle and r its radius. The radius of the circle, using the distance formula, equals: $r = \sqrt{(m - h)^2 + (n - k)^2}$. Hence, the desired equation is: $(x - h)^2 + (y - k)^2 = (m - h)^2 + (n - k)^2$.

90. (Fig. 270) The radius of circle C is : $r = \sqrt{(a - h)^2 + (b - k)^2}$ and $r^2 =$ $(a - h)^2 + (b - k)^2$. The desired equation is: $(x - h)^2 + (y - k)^2 = (a - h)^2 + (b - k)^2$.

91.

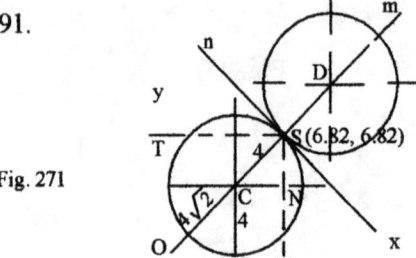

Fig. 271

In the figure, $CN = SN = 4 / \sqrt{2}$ $= 2 \sqrt{2} = 2.82$. (CNS is a 45-45 rt $\triangle$, and each leg = hyp$/\sqrt{2}$). Thus, $SM = 4 + 2 \sqrt{2} = 6.82$. The slope of m = 1, hence the slope of n $= -1$. Thus, the equation of the line n is: $y - 6.82 = -1 (x - 6.82)$ or $y = -x + 13.64$.

92.

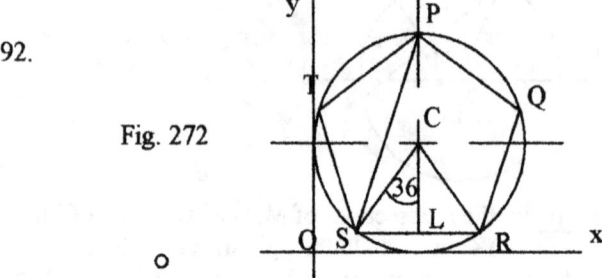

Fig. 272

Each central angle of the pentagon = 360 / 5 = 72 deg. Thus, $\angle SCL$ $= 36°$. SL = 6 sin 36 = 3.54; CL = 6 cos 36 = 4.86. S is at (2.46, 1.14). From the figure, the vertex P is at (6, 12). Hence, the equation of SP is: $y - 12 = [(12 - 1.14) / (6 - 2.46)] (x - 2.46)$ or $y = 3.06x + 4.47$.

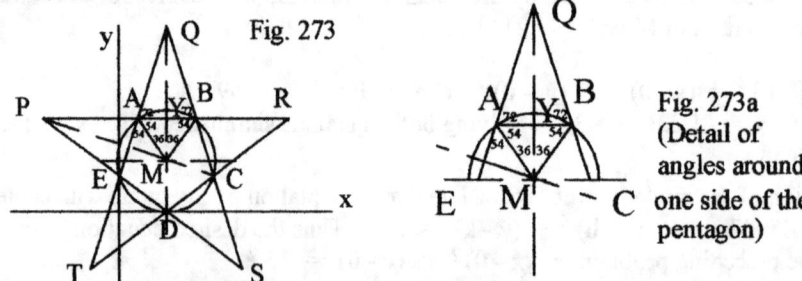

Fig. 273

Fig. 273a (Detail of angles around one side of the pentagon)

93. The figure consists of (1) the inside regular pentagon, and (2) the 5 triangles around the pentagon. Each central angle of the pentagon = 360 / 5 = 72 deg. (See Fig. 273a) Thus $\angle AMY = 36°$. AY = s / 2 = 8 sin 36 = 4.72. So s = 9.44. A of pentagon = (1 / 2) aP = (1/2) 8 cos 36 (5) (9.44) = 152.93 sq. units.

Area of the 5 triangles = 5 [1 / 2 (AB) YZ)] = 5 [1 / 2 (9.44) (AY tan 72)] = 5 [1 / 2 (9.44) (4.72 x 3.08)] = 342.91 . Hence, area of the 5-point star = area of the pentagon + area of the 5 triangles = 153.93 + 342.91 = 495.84 sq. in

94.

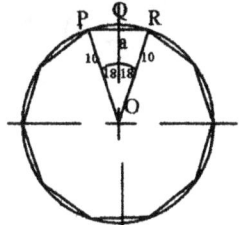

Fig. 274

Each central angle of the decagon, such as∠ PQR = 360 / 10 = 36 . Likewise, ∠ POQ = 18 . Thus s / 2 = PQ = 10 sin 18 = 3.1, so s = 6.20. Thus A = 1 / 2 a P = 1 / 2 (10 cos 18) (10 x 6.20) = 294.5 sq. in.

95.

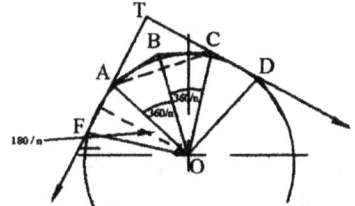

Each central angle of the n-gon, such as∠ AOB = 360 / n , so ∠ AOC = 720 / n. The desired measure of∠ T = 180 - (∠ TAC +∠ TCA), where ∠ TAC = ∠ TCA = { 180 - [90 - (360 / n) + 80 - (180 / n)] } = 540 n; Hence T = 180 - 2∠ TAC = 180 - (1080 / n) = (180n - 1080) / n.

96.

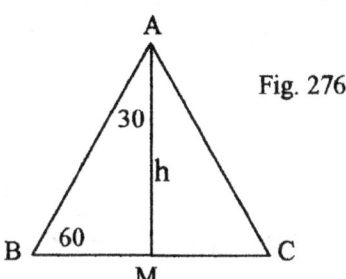

Fig. 276

The area of an equilateral triangle with side s is: $A = (s^2 / 4)\sqrt{3}$. $A = (s^2 - 12x + 36)\sqrt{3} = (s^2 / 4)\sqrt{3}$. So s = 12, s / 2 = 6, and the altitude $h = 6\sqrt{3}$. In a 30 - 60 rt △ , the longer leg equals the shorter leg x $\sqrt{3}$.

97.

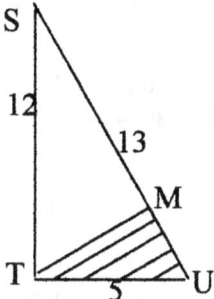

By similar triangles, TM / ST = UT / US, TM / 12 = 5 / 13; TM = 60 / 13 = 4.62. UM / UT = UT / ST; UM / 5 = 5 / 12, and UM = 25 / 12 = 2.08. Area of △UMT = 1 / 2 (UM) TM = 5.66 sq. in.

98.

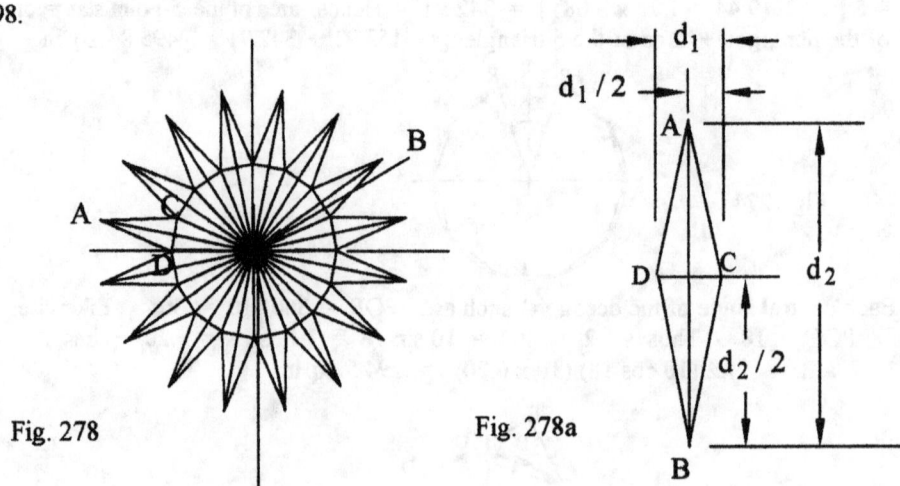

Fig. 278   Fig. 278a

Fig. 278a shows that the area of a rhombus is the combined areas of 4 congruent right triangles. The area A of the rhombus is: $A = 4(1/2)(d_1/2)(d_2/2) = (1/2) d_1 d_2$.

Fig. 278 shows that the sunflower is made up of 16 congruent rhombuses whose diagonals d_1 and d_2 are DC and AB respectively.
Clearly, the area of the sunflower = $16(1/2)$ AB x CD or 8 AB (CD).

99.

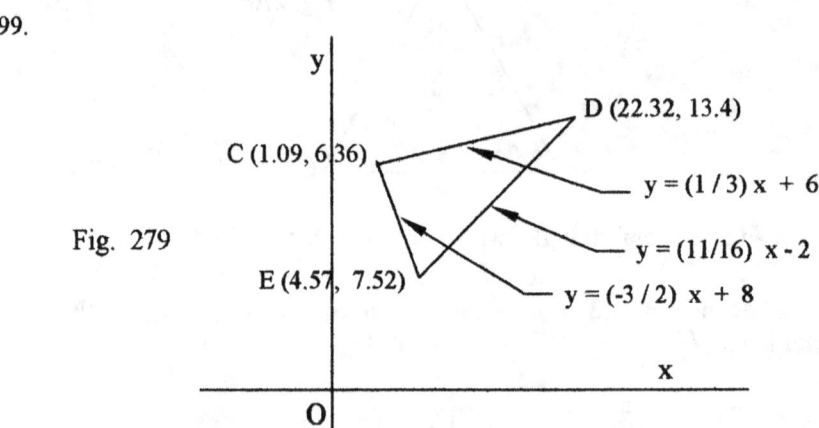

Fig. 279

The three lines intersect to form triangle CDE, with vertices at the given coordinates. The vertices are the intersection points of the lines, obtained by solving the equations of the lines simultaneously.

Find the lengths of the sides, using the distance formula, and then use Hero's Formula to determine the area.

To locate point C: Solve both equations below to find the intersection points:
                                $y = (1/3) x + 6$; $y = (-3/2) x + 8$
                                $x = 1.09$ and $y = 6.36$

To locate point D: Solve both equations below to find the intersection points:
                                $y = (1/3) x + 6$; $y = (11/16) x - 2$
                                $x = 22.32$, $y = 13.41$

To locate E:

$y = (-3/2)x + 8$; $y = (11/16)x - 2$; Solving the system, $x = 4.57$, $y = 7.52$.

To find the lengths of the sides of the triangle:

$CD = \sqrt{(22.22 - 1.09)^2 + (13.41 - 6.36)^2} = 20.22$

$DE = \sqrt{(22.22 - 4.59)^2 + (10.41 - 7.52)^2} = 16.63$

$CE = \sqrt{(4.57 - 1.09)^2 + (7.52 - 6.36)^2} = 11.58$

Using Hero's Formula,

$A = \sqrt{s(s-a)(s-b)(s-c)}$, where $s = (a+b+c)/2 = (20.22 + 16.63 + 11.58)/2 = 24.11$.

$= \sqrt{24.11(4.09)\,13.48\,(12.53)}$

$= 129.06$

100.

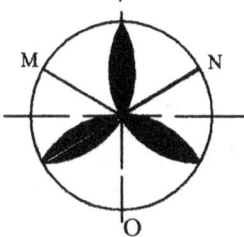

Fig. 280

Each of the three "leaves" in the cluster is made up of 2 segments (a segment of a circle is the portion left after removing the triangular area in a sector) from two adjacent sectors rotated to face each other and form a "leaf" as shown in Fig. 230.

$A_{3\text{ leaves}}$ = area of 6 segments

$= 6\,(A_{\text{sector}} - A_{\triangle})$

$= 6\,[\,(60/360)\,\pi 10^2 - (s^2/4)\,\sqrt{3}\,]$

$= 54.66$ sq. in.

101. The probability P of the parachutist landing on the park situated on the island is equal to the ratio of the area of the park to the area of the island.

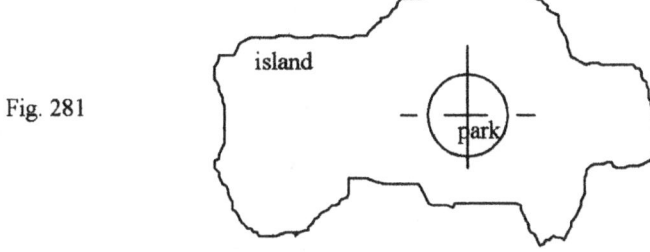

Fig. 281

P (landing on the park) $= A_{\text{park}} / A_{\text{island}}$

$= [\,3.14\,(.25)^2 / 10\,] = .02$

Solution 1:

1. Extend the upper base b to M and N ; draw altitudes PM and ON to form the two right triangles PMS and QNR.

2. From the figure, the area of the trapezoid equals the area of the rectangle PMNO + the area of the two triangles PMS and QNR.

$A = ah - (1/2)hx - (1/2)h(a-b-x) = (1/2)h(a+b)$.

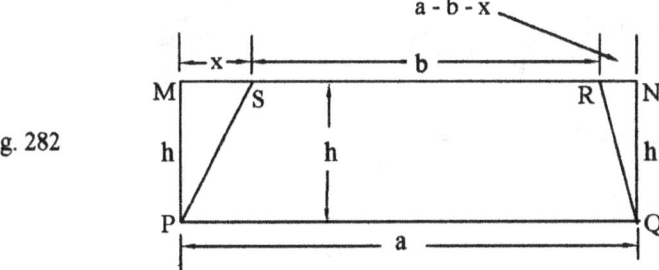

Fig. 282

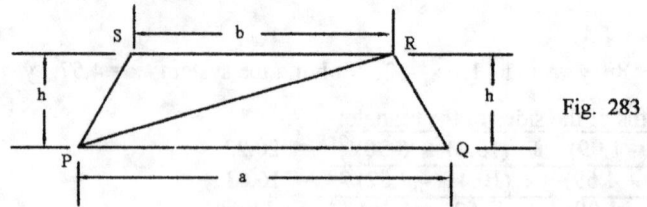

Fig. 283

Solution 2. 1. $A_{trap} = (1/2)\ ah$ 2. $A_{trap} = (1/2)\ bh$
3. $A_{trap} = (1/2)\ h\ (a + b)$ (Adding 1 and 2).

103.

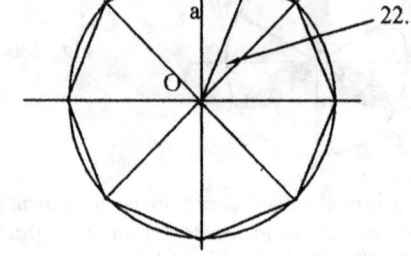

Fig. 284

To solve for s and a :

$a / 10 = \cos 22.5$
$a = 10 \cos 22.5 = 9.2$

$(s / 2) / 10 = \sin 22.5$
$(s / 2) = 10 \sin 22.5$
$s = 15.32$

$A_{segment} = A_{sector} - A_{\triangle}$
$= [(45/360)\ \pi\ 10^2 - (1/2)\ s\ a = [(1/8)\ 3.14\ (100) - (1/2)\ 15.32\ (9.2)$
$= 4.01.$ Hence, area of 8 segments $= 8\ (4.01) = 32.08$; Area of 8 sectors $= 8\ (45/360)\ \pi(10)^2$
$= 314$ $A_{segment} / A_{sector} = 32.08 / 314 = .10$

104.

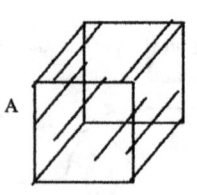

A

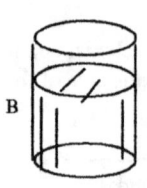

B

Fig. 285

Volume of water pumped from A must equal the volume of water contained in B.
Vol. of the cube = vol of the cylinder. $8^3 = \pi r^2\ (8)$; $512 = 3.14\ r^2\ (8)$;
$r^2 = 512 / 25.\ 12 = 20.39$; $r = 4.51$; $d = 9.02$ ft.

105.

Fig. 286

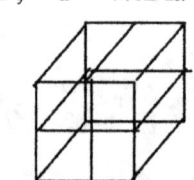

The lead in the cube must equal the lead in the cylinder.
Vol of the cube = vol of the cylinder
$10^3 = \pi\ r^2\ (4)$; $1000 = 3.14\ (4)^2\ h$; $h = 19.90$ in.

Fig. 287

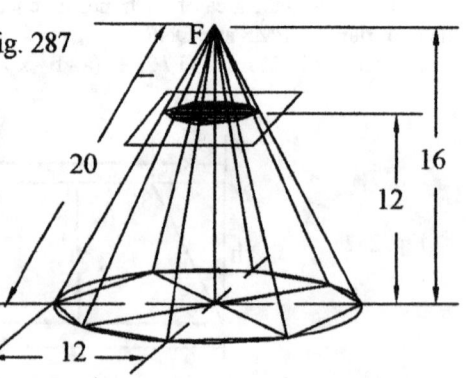

106. The frustum of a pyramid is the
portion of the pyramid left after its top
is cut off by a cutting plane parallel to
its base.

Thus the volume V of the frustum is

$V_{frustum} = V_{big\ pyramid} - V_{small\ pyramid}$

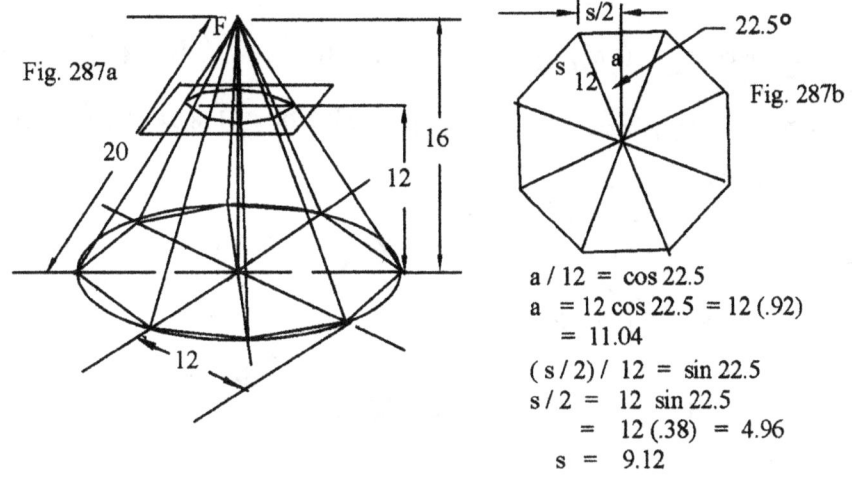

Fig. 287a

Fig. 287b

$a / 12 = \cos 22.5$
$a = 12 \cos 22.5 = 12 (.92)$
$= 11.04$
$(s / 2) / 12 = \sin 22.5$
$s / 2 = 12 \sin 22.5$
$= 12 (.38) = 4.96$
$s = 9.12$

Vol of pyramid $= (1 / 3)$ A of base x height $= (1 / 3) [(1 / 2) a P] $ x h
$= (1 / 3) [(1 / 2) 11.04 (8 \times 9.12)] $ x $ 16 = 2147.94$ cu. in.

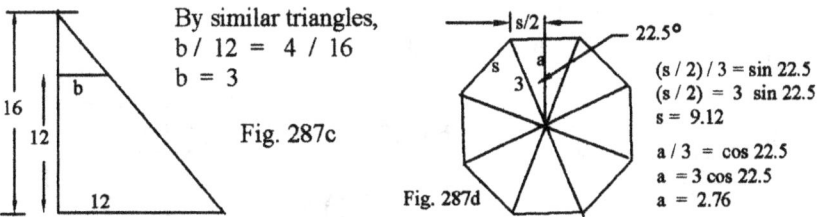

By similar triangles,
$b / 12 = 4 / 16$
$b = 3$

Fig. 287c

Fig. 287d

$(s / 2) / 3 = \sin 22.5$
$(s / 2) = 3 \sin 22.5$
$s = 9.12$

$a / 3 = \cos 22.5$
$a = 3 \cos 22.5$
$a = 2.76$

Vol of small pyramid $= (1/3) (1/2 $ a P$) $ h $= (1/3) (1/2) 2.76 (8) 2.24 \times 4 = 34.23$ cu. in.
Vol of frustum = Vol of big pyramid - V small pyramid $= 2147.94 - 34.23$
$= 2113.71$ cu. in.

107.

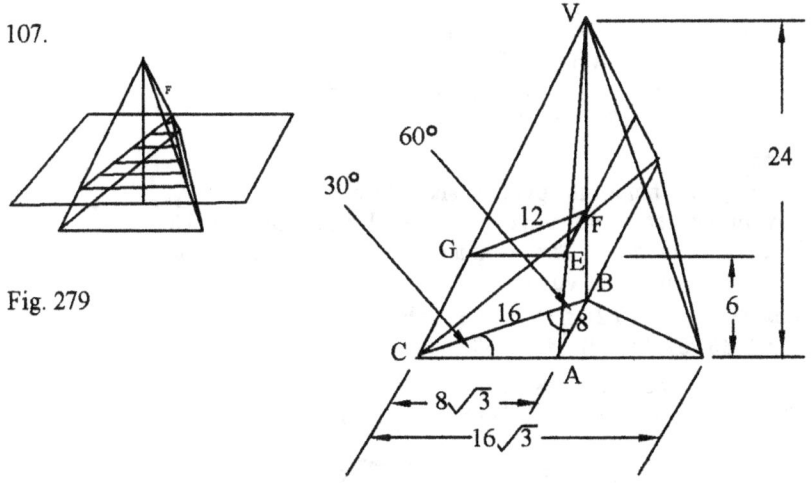

Fig. 279

In rt $\triangle$CAB, $\angle$BCA $= 30°$, $\angle$CBA $= 60°$, CA $= 8\sqrt{3}$, so AB $= 8\sqrt{3} / \sqrt{3} = 8$, BC $= 16$.

RT $\triangle$ VGF $\sim$ rt $\triangle$ VGB; by similar triangles, CF / CB = VF / VB,
GF / 16 = 18 / 24, GF = 12.

In rt $\triangle$ GEF, GF = 12, EF = 6, so GE = s / 2 = $6\sqrt{3}$, and s = $12\sqrt{3}$.

Vol of small pyramid:
V $= (1 / 3) $ A $_{base}$ x h, where A $_{equilateral \triangle} = (s^2 / 4) \sqrt{3}$.
$= (1 / 3) [(12\sqrt{3})^2 / 4 (\sqrt{3}) 18] = 1118.04$ cu. in.

Vol of big pyramid = (1 / 3) (A   h) = (1 / 3) { [(16   3) / 4] }   3} 24
= 2657.28 cu. in.

$V_{frustum}$ = $V_{big\ pyramid}$ - $V_{small\ pyramid}$ = 2657.28 - 1118.04 = 1539.24 cu.in.

Fig. 280

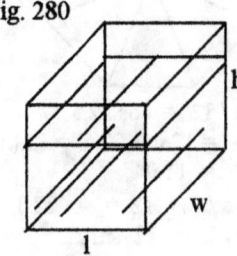

Fig. 280a

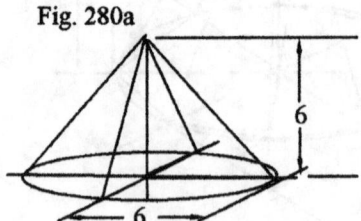

108. The volume of the cone must equal the volume of the molten lead.

V_{cone} = V_{lead}

(1 / 3) A_{base} x h = l x w x h

(1 / 3) $\pi\ 6^2$ (6) = 10 x 12 x h ; h = 1.88

109.

V_{sphere} = $V_{water\ overflow}$

(4 / 3) $\pi\ r^3$ = 22 (22) h

(4 / 3) $\pi\ (10)^3$ = 484 h ; h = 8.65 in.

110.

Fig. 281

Fig. 281a

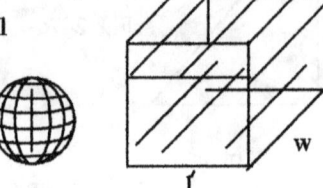

Fig. 282

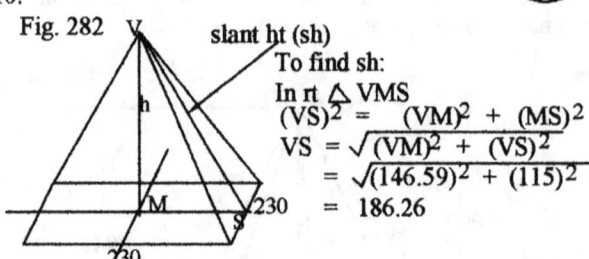

slant ht (sh)

To find sh:

In rt $\triangle$ VMS

$(VS)^2$ = $(VM)^2$ + $(MS)^2$

VS = $\sqrt{(VM)^2 + (VS)^2}$

= $\sqrt{(146.59)^2 + (115)^2}$

= 186.26

The total covering material for the pyramid should have
The total covering material for the pyramid should have an area equal to
the lateral area of the pyramid. The amount of concrete needed to duplicate the structure is equal to its volume.

LA = Lateral Area = (1 / 2) slant ht x P of base

    = (1 / 2) (186.26) 4 (230) = 85697.60 sq. m.

Vol of concrete = (1 / 3) A of base x h

= (1 / 3) (240 x 240) 146.59 = 2,584,870.33 cu. m.

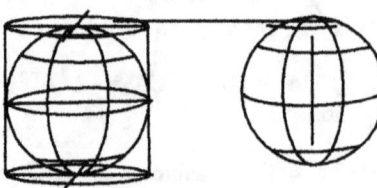

Fig. 282

111. Vol of water poured before overflow = V of cyl - V of sphere

Vol of cyl = $\pi\ r^2$ h = 3.14 (6)^2 (12) = 1356.48

Vol of sphere = (4 / 3) $\pi\ r^3$ = (4 / 3) 3.14 (6)^3 = 902.06

Vol of water poured before overflow = 1356.48 - 902.06

    = 454.42 cu. in.

112. At K, draw a line m which forms any size angle with Kl. Using a compass, draw 10 equally spaced arcs to form 9 equal divisions on line m. Locate points P, Q, and R on the ist, 3rd, and 9th divisions on m. PK, QK, and RK are the desired segments.

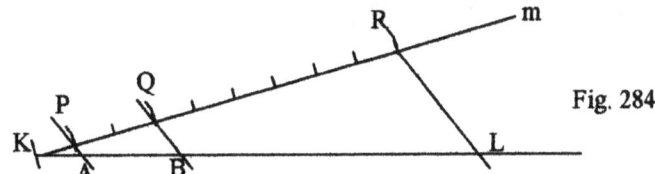

Fig. 284

113. Draw a working line w as shown. At some point O on the line, copy ∠M twice, each adjacent to the other. Adjacent to side OA, copy ∠N. Bisect ∠BOC, then ∠ DOC is the desired angle.

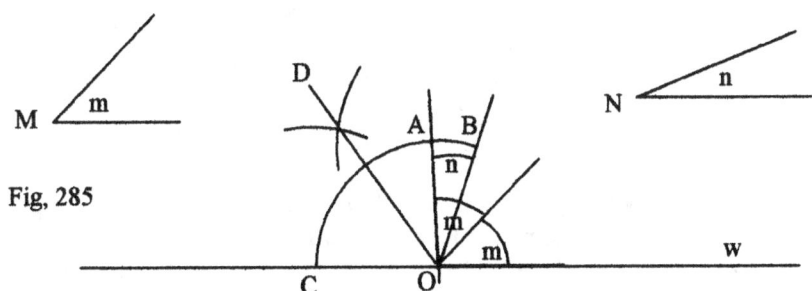

Fig, 285

114.

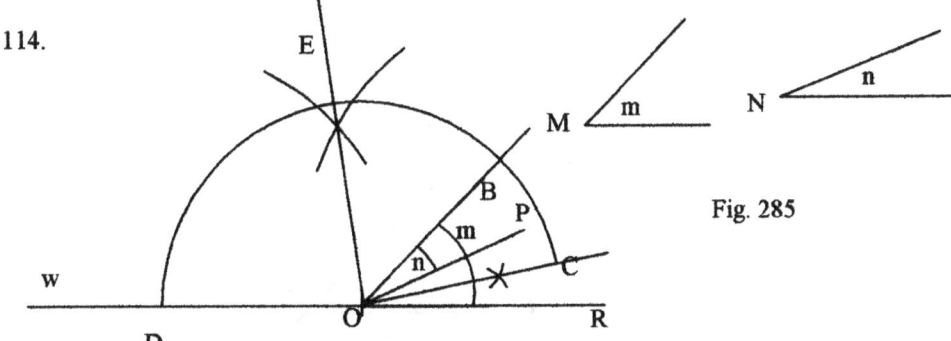

Fig. 285

Draw working line w. At some point O, copy ∠M. Adjacent to side OB, copy ∠N. Bisect ∠POR to get the angle (m - n) / 2 with a side at OC. Bisect the angle COD to obtain OE.. ∠DOE is the desired angle.

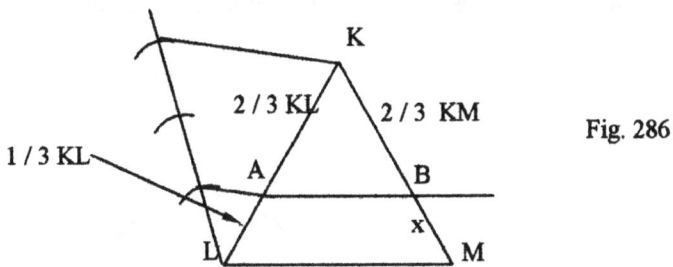

Fig. 286

115. 1. Divide the left side KL into 3 equal segments as shown. 2. Draw line AB parallel to LM. 3. (2 / 3) KL / (1 / 3) KL = (2 / 3) KM / x. (A line which intersects one side of a triangle and is parallel to one side divides the other side into segments proportional to segments on the first side.

4. Solving the proportion for x, x = [(1 / 3) KL (2 / 3) KM] / (1 / 3) KL] .
5. Divide x, as in the example given, into three parts to obtain the desired segment.

116.

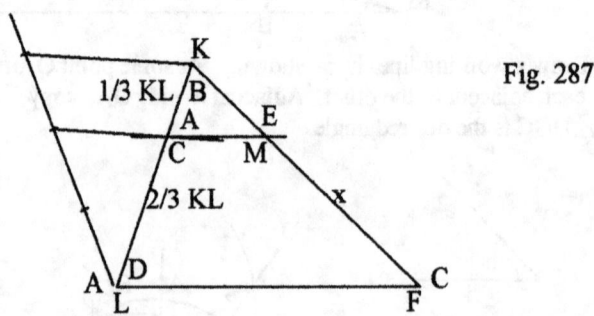

Fig. 287

1. From the figure, KC = (1 / 3) KL and CD = (2 / 3) KL. 2. At K, draw KE = BM = (1 / 3) KF.   3. Connect C to E, and draw DF II CE.   4. Since DF IICE, then (1 / 3) KL / (2 / 3) KL = (1 / 3) KF / x (A line parallel to one side of a triangle divides the other two sides into proportional segments).

117.

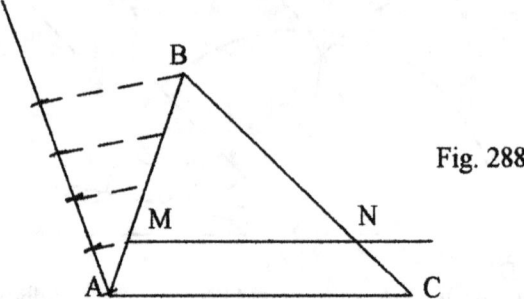

Fig. 288

1. As in the preceding two problems, divide AB into four equal segments.   2. At M, draw MN II AC.   3. Then AM / MB = CN / NB. (A line intersecting one side of a triangle and parallel to one side, divides the other side into segments proportion to the segments of the first side.

118.

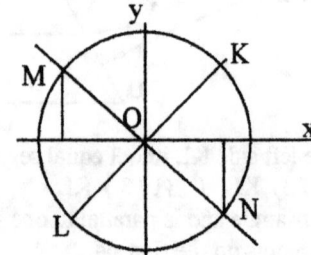

Fig. 289

1. In Fig. 289,   MRO and   NSO are 45 - 45 right triangles (MN   KL is given) .
2. MR = 10 sin 45 = 7.10.   OR = 10 cos 45 = 7.10, hence the coordinates of M are
(-7.10, 7.10)   3. NS = 10 sin 45 = 7.10.   OS = 10 cos 45 = 7.10. Hence, the
coordinates of N are (7.10, -7.10).

Fig, 290

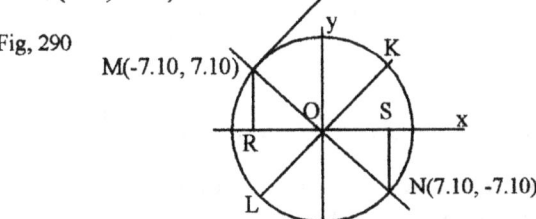

119.   The slope of MN is -1 , hence the slope of the tangent (perpendicular to the radius of
the circle at the poin of tangency) is 1.   Therefore, the equation of the tangent is
y - 7.10 = (1) (x + 7.10) , or   y = x + 14.20.

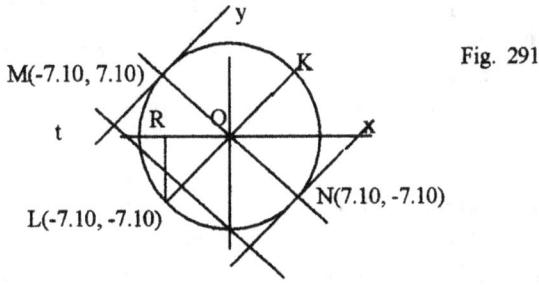

Fig. 291

120.

1. In Fig. 291, the slope of the tangent at N is 1, so the slope of the line perpendicular to this
tangent is -1.   2. ORL is a 45 - 45 right triangle, thus   RL = 10 sin 45 = 7.10, and OR = 10
sin 45 = 7.10.   3. Therefore, the equation of the line (tangent t) through L perpendicular
to the tangent at N is: y + 7.10 = - 1 (x + 7.10), or y = -x - 14.20.

121. 1. Draw a working line w.   2. Copy, on w, PQ twice to obtain PR, and RS once.
3. Bisect segment PS to locate its midpoint M, and with M as center, draw a half circle passing
through P and S.
4. Through S, draw an altitude RF of right triangle PFS.   5. PR / RF = RF / RS (the
altitude to the hypotenuse of a right triangle is the geometric mean between the segments - PR
and RS - of the hypotenuse).   6. The altitude RF, therefore, is the desired segment.

R ————————— S

P ——————— Q

Justification:

PR / RF = RF / RS

$(RF)^2$ = PR (RS)

RF $= \sqrt{PR(RS)}$

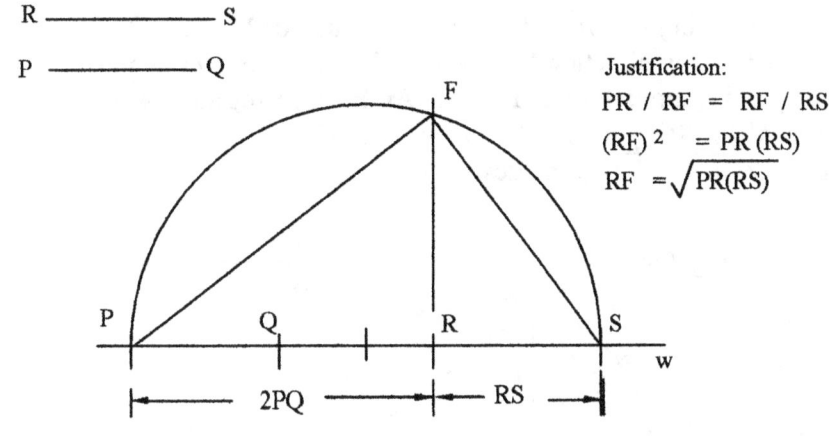

122.

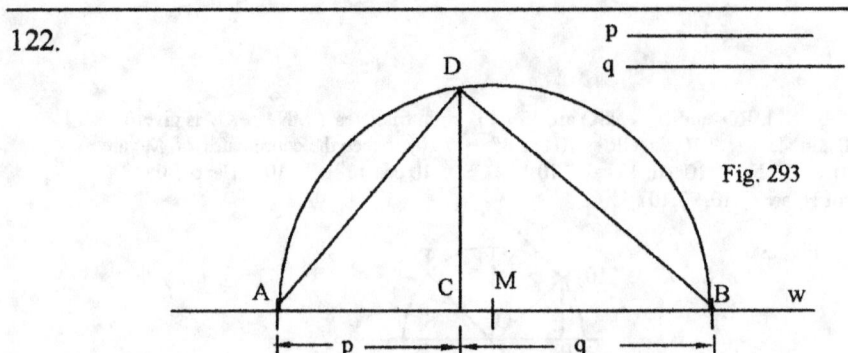

Fig. 293

1. Draw working line w and copy segments p and q end to end on w as shown. 2. Bisect AB to locate the midpoint M, and with M as center, draw a half-circle through A and B. 3. Through C, draw altitude CD and connect A to D and B to D to draw right triangle ADB.
4. AC / CD = CD / BC, or p / CD = CD / q, or $(CD)^2 = pq$. (The altitude to the hypotenuse of a right triangle is the geometric mean between the segments of the hypotenuse.
5. Hence, CD = $\sqrt{pq}$. Copy $\sqrt{pq}$ 3 times to obtain 3 $\sqrt{pq}$, the desired segment.

123. 1. Draw working line w. 2. On w, draw segments equal to (1/2) q and to 5p; position them end to end as shown in Fig. 294. 3. Bisect RS to locate M, then draw a half-circle which passes through R and S. 4. At U, draw altitude UV of right triangle RVU. 5. RU / UV = UV / US (The altitude to the hypotenuse of a right triangle is the geometric mean between the segments of the hypotenuse). 6. UV = $\sqrt{RU\,(US)}$ = $\sqrt{(1/2)\,q\,(5p)}$ = $\sqrt{(5/2)\,pq}$, the desired segment.

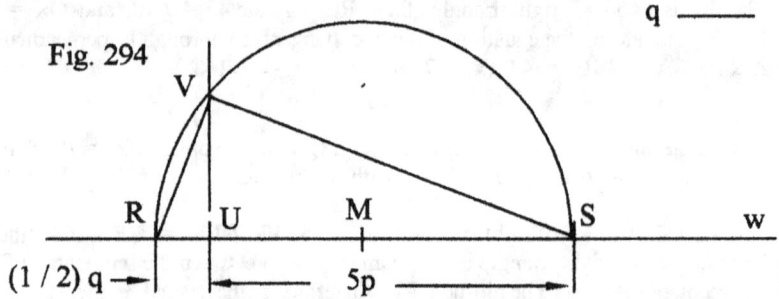

Fig. 294

124. 1. On working line w, draw segments equal to 2AB and 3PQ. 2. Locate center C by bisecting MN, and with C as center, draw the half-circle passing through M and N. 3. At P, draw the altitude h of right triangle MSN. 4. 2AB / h = h / 3PQ.
Thus h = $\sqrt{2AB\,(3PQ)}$, the desired segment.

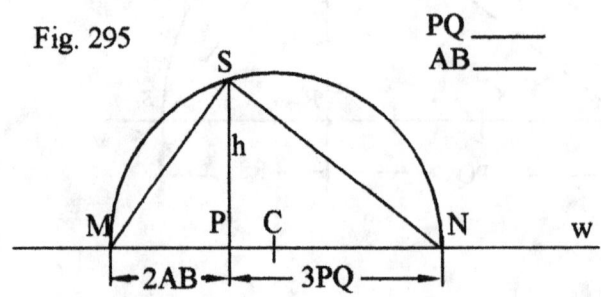

Fig. 295

125.

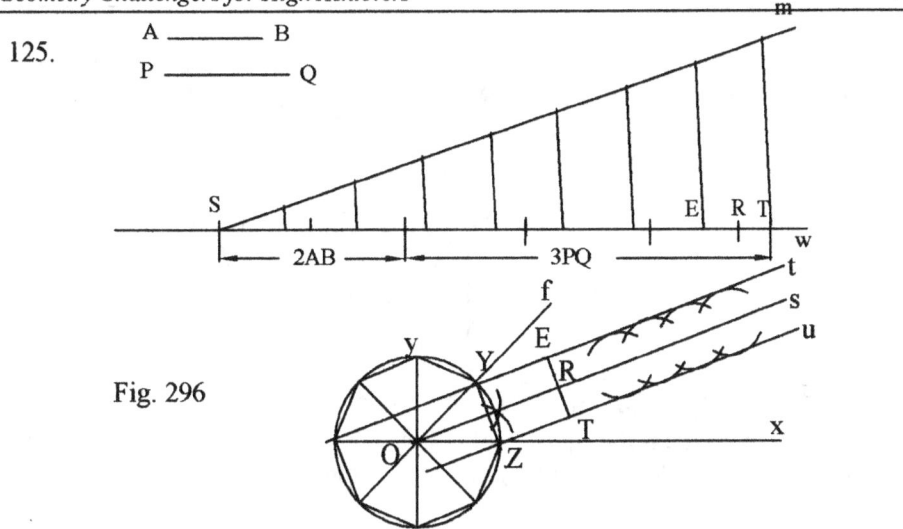

Fig. 296

1. Draw a working line w and on it, draw end to end, segments equal to 2AB + 3PQ.
2. Draw an auxiliary line m, and on it, draw 8 equal segments as shown using a compass. 3. Draw 8 parallel lines from m to segment ST on w. 4. ET is 1/8 of ST; bisect ET to obtain midpoint R. 5. Draw the x and y axes, bisect each of the four angles they form; bisect also ∠fox with bisector s. 6. At 4 random points on s, draw arcs each with RE as radius. 7. On both sides of s, draw tangents l and t to these arcs, the tangents intersecting the x-axis at Z and the angle bisector at Y. 8. Connect Y to Z, then YZ is one side of the required octagon. 9. Draw a circle with center at O, and with radius OY or OZ and connect the intersection points the circle makes with the rest of the angle bisectors to form the required octagon.

126. The centroid of a triangle is the point of intersection of its medians. 1. Bisect each side of the triangle in Fig. 287 to locate the midpoints A, B, and C, then connect each point to vertices S, T, and O as shown. 2. The intersection point K is the desired centroid with coordinates (1.67, 4).

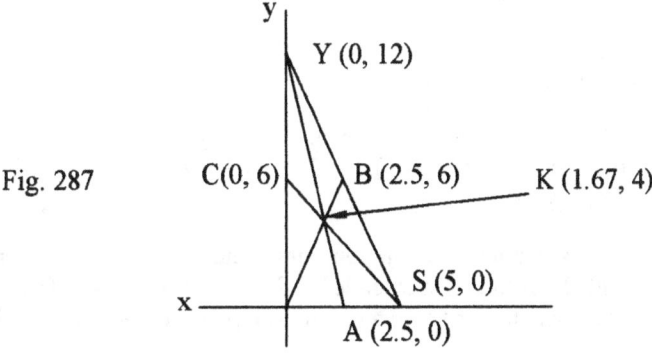

Fig. 287

Eq. of TA: y - 12 = [(12 - 0) / 0 - 2.5)] (x - 0) or y - 12 = -4.8x

Eq. of SC: y - 0 = [(6 - 0) / (0 - 5)] (x - 5) or y - 0 = -1.2x + 6

Solving both equations simultaneously: x = 1.67, y = 4. The centroid is at (1.67, 4).

Obviously, the median OB must pass through K(1.67, 4). To check, show that the slope of OK is equal to the slope of KB.

127.

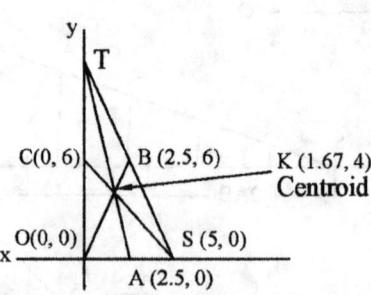

Eq. of OB: y - 0 = [(6 - 0) / (2.5 - 0)] (x - 0), or y = 2.4x
Eq. of OK: y - 0 = [(4 - 0) / (1.67 - 0)] (x - 0) or y = 2.4x
The desired equation is the equation of OK which is y = 2.4x, same as the equation of OB.   This is true since O, C, and B are collinear.

128.

1. The circumcenter of a triangle is the center of the circumscribed circle, and is determined by the intersection of the perpendicular bisectors of the sides of the triangle.   2. Draw the perpendicular bisectors d, e, and f.   3. As shown, the bisectors intersect at S(2.5, 6). (Note that the circumcenter of any right triangle is the midpoint of the hypotenuse).

Fig. 298

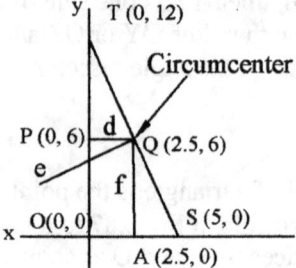

To locate the circumcenter analytically , find the intersection of the two perpendicular bisectors

     Eq. of f: s = 2.5
     The intersection is at (2.5, 6).

129. (Refer to Fig. 298, prob. 128). The circumcenter is Q(2.5, 6).
Eq. of OQ: y - 6 = [(6 - 0) / (2.5 - 0)] (x - 2.5), or y = 2.4x
d and f: Eq. of d: y = 6

130. The orthocenter is the intersection point of the altitudes. Clearly, from the figure, the altitudes OT and OS of the right triangle TOS intersect at O(0, 0). This indicates that for all right triangles, the orthocenter is the vertex of the right angle in the right triangle.

Fig. 299

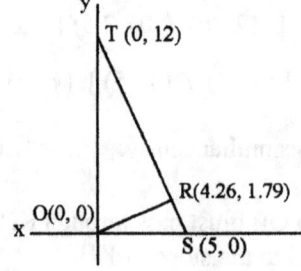

131. The origin is of course at (0, 0), but the orthocenter is also at (0, 0), as seen in prob. 130. Therefore, the equation of the line from the origin to the orthocenter is: x = 0, y = 0

132.

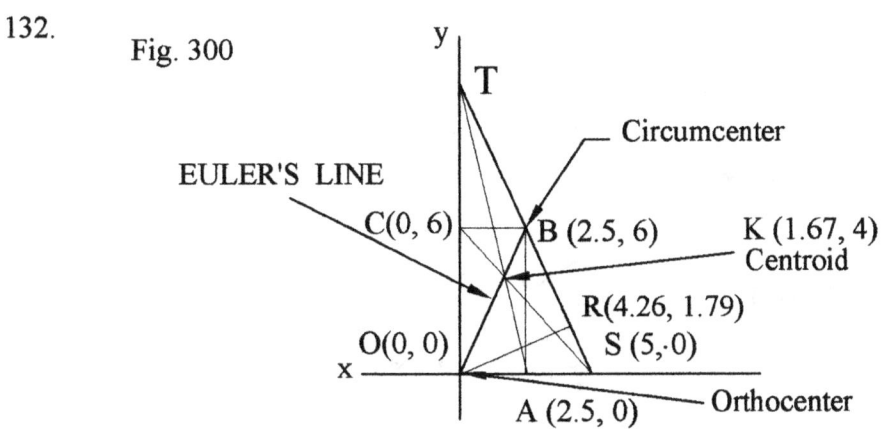

Fig. 300

Eq. of Euler's Line: y - 4 = [(6 - 4) / (2.5 - 1.67)] (x - 1.67) or y = 2.4x
            also,   y - 6 = [(6 - 0) / (2.5 - 0)] (x - 2.5)   or y = 2.4x

133.

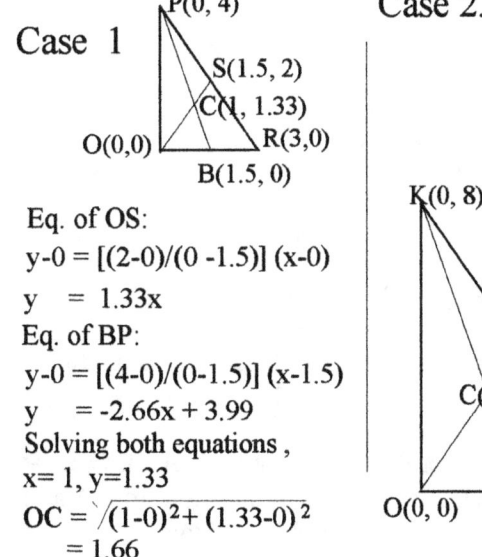

Case 1

Eq. of OS:
y-0 = [(2-0)/(0 -1.5)] (x-0)
y   = 1.33x
Eq. of BP:
y-0 = [(4-0)/(0-1.5)] (x-1.5)
y   = -2.66x + 3.99
Solving both equations ,
x= 1, y=1.33
OC = $\sqrt{(1-0)^2 + (1.33-0)^2}$
      = 1.66

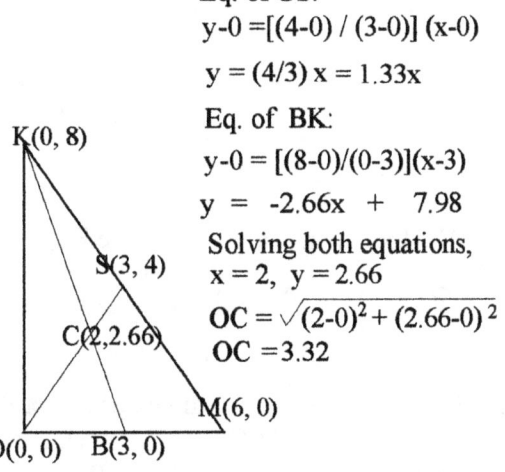

Case 2. Case 1 dimensions doubled

Eq. of OS:
y-0 =[(4-0) / (3-0)] (x-0)

y = (4/3) x = 1.33x

Eq. of BK:
y-0 = [(8-0)/(0-3)](x-3)

y = -2.66x + 7.98

Solving both equations,
x = 2, y = 2.66

OC = $\sqrt{(2-0)^2 + (2.66-0)^2}$
OC =3.32

The above solution shows that when the dimensions of the given triangle are doubled, the distance of the centroid C from O is also doubled (the distance increases by 100%).

134. From the figure, the locus is the frustum of a cone (the portion left after the top is cut off by a cutting plane parallel to the base).

Fig. 302

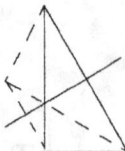

135. The locus is a "donut" (a circular tube) with outside radius 10" and inside radius 4".

Fig. 303

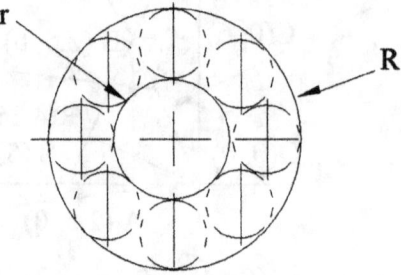

136. The locus is a circle with a greater radius R. From the figure, note that

the tangents to the circle intersect outside the circle at such points as 1 and 2 as shown in Fig. 304. The intersection points are all at a distance R from the center of the circle, forming a larger circle with radius R.

In the figure, r / R = cos 9, hence r = .98R.

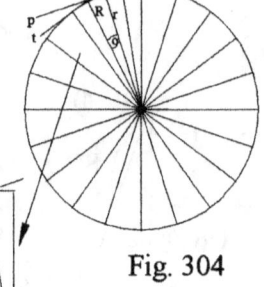

Fig. 304

137. The locus is a straight tube (see Fig. 305) with a diameter of 6".

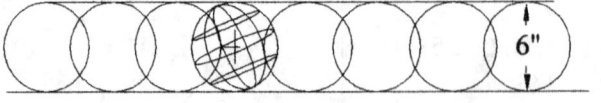

6" Fig. 305

138. There are two loci: 1) the locus of points equidistant from A, B, C, D, and E is the line n in Fig. 306; and 2) the locus of points equidistant from A and B is the plane halfway between A and B. The two loci do not intersect since n lies on plane P.

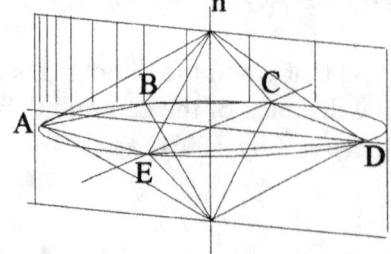

Fig. 306

139. The locus is the plane that is coplanar with the perpendicular bisector of PQ. The two loci do not intersect, since line WY lies on the plane.

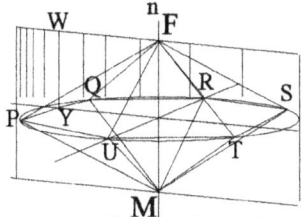

Fig. 307

140. The locus of points equidistant from R and U is the same plane that is coplanar with the perpendicular bisector MN of PQ.

Fig. 308

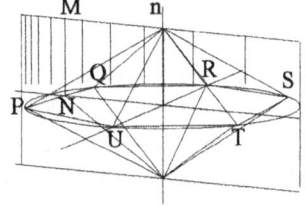

141. Fig. 309 shows that: KR = KS = KT, OR = OS = OT, and LR = LS = LT. The locus, therefore, of points equidistant from R, S, and T is the line n which is perpendicular to the plane of the hexagon.

Fig. 309

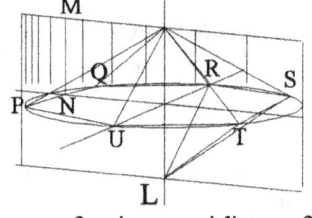

142. There are three loci, namely: The locus of points equidistant from Q and R is the bisecting plane q of the segment QR; 2. the locus of points equidistant from S and T is the bisecting plane q of the segment ST; and 3. the locus of points equidistant from R and U is the bisecting plane p of the segment U (incidentally identical to locus 2).

Fig. 310

143.

Fig. 311

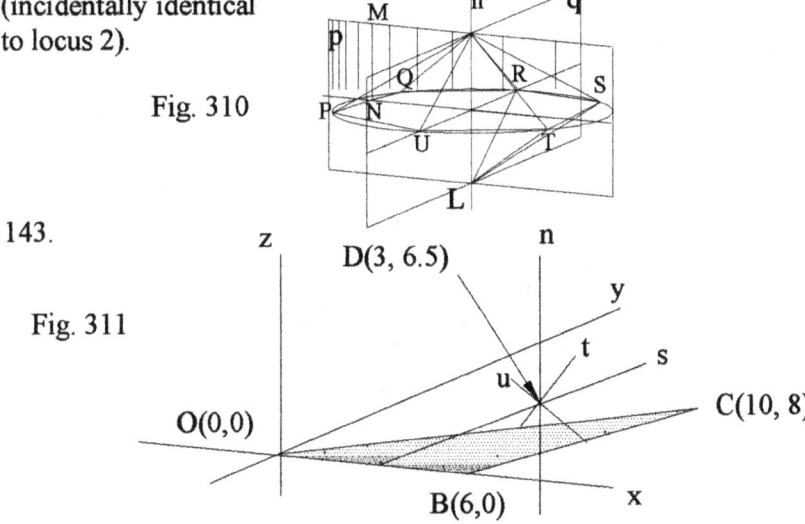

The desired locus is the line n, which passes through the circumcenter D, and is perpendicular to the plane of the triangle. All points on the line n are equidistant to O, B, and C. D is the intersection of the perpendicular bisectors of the sides. To locate D: Eq. of s: $x = 3$; Eq. of t: (slope of Oc = 4/5, hence slope of t = -5/4) $y-4 = (-5/4)(x-5)$. Solving the equations simultaneously, $x = 3$, $y = 6.5$. D is (3, 6.5). D is where the line n intersects the x-y plane.

144.

Fig. 312

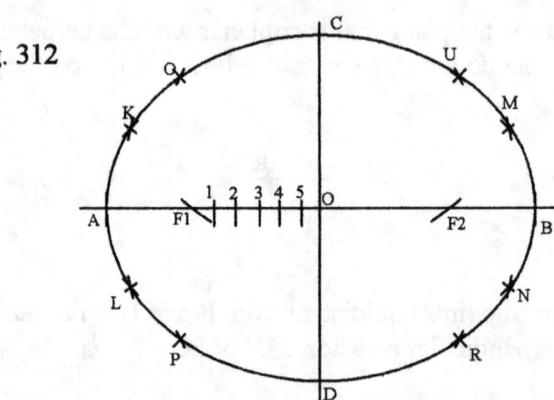

An ellipse can be constructed by the definition method, a method based precisely on the definition of an ellipse. Definition: An ellipse is the set of all points on a plane such that the distance of a point to F1 and F2 added together equals a constant k (where k is the length of the major diameter AB). F1 and F2 are fixed points on the major axis AB and are called the foci.

To construct the ellipse:

1) With AO as radius, and C as center, draw two arcs as shown intersecting AB at F1 and F2.

2) Pick 5 random points between F1 and O, and number them 1, 2, 3, 4, and 5.

3) With A1 as radius and F1 and F2 as centers, draw arcs at K, L, M, and N.

4). With B1 as radius and F1 and F2 as centers, draw arcs at K, L, M, and N.

5). With A2 as radius and F1 and F2 as centers, draw arcs at O, P, Q, and R.

6) With B2 as radius, and F1 and F2 as centers, draw arcs at O, P, Q, and R.

7) Similarly, use A3-B3, A4-B4, and A5-B5 as radii and draw 6 more points.

8) Connect the points to draw this ellipse.

148. Definition: A parabola is the set of all points on a plane such that the distance of every point to a fixed line called the directrix is equal to its distance from a fixed point called the focus. In the figure, V, or vertex, is halfway between the directrix and the focus.

Fig. 313

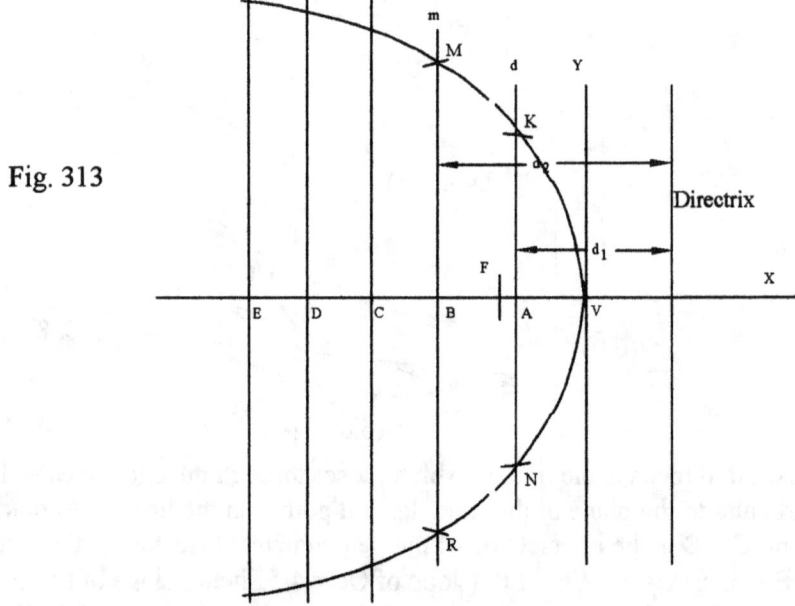

1) Draw d perpendicular to the x-axis through random point A.

2) Set compass to d_1, and with d_1 as radius and F_1 as center, draw arcs intersecting d at K and L.

3) Draw m perpendicular to the x-axis through random point B.

4) Set compass to d_2, with d_2 as radius and F_2 as center, draw arcs intersecting m at M and N.

5) Repeat steps 1 to 4 at points C, D, and E

6) Connect all points to draw the parabola.

147. Problems 147 to 150 are exercises in describing shapes of objects from various positions. They are related to construction and loci problems.

1. Through points B, A and C, D, draw vertical projection lines l and m intersecting the 45-deg line (called mitre line) w at S and T. 2. Through S and T, draw horizontal (projection) lines n and o to intersect vertical (projection) lines q and p at points A, B, C and D. 3. ABCD is the desired top view.

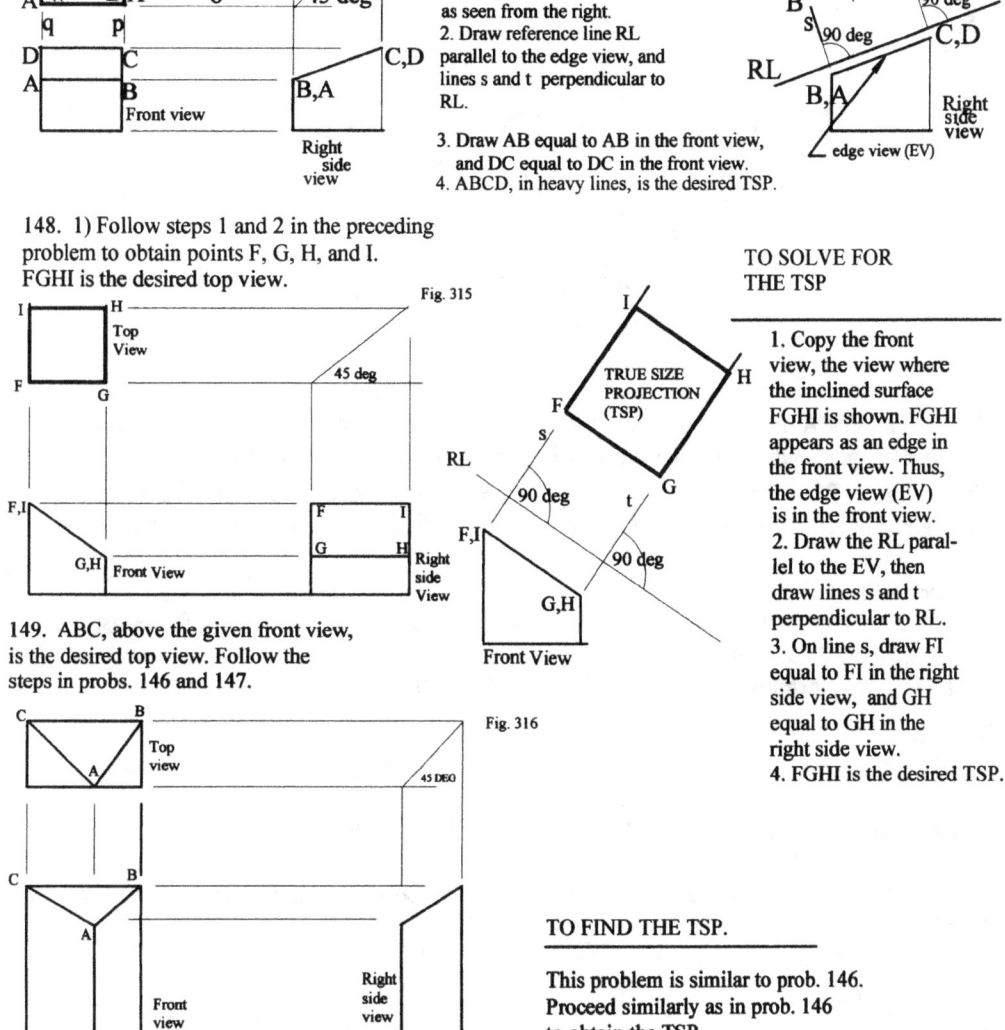

Fig. 314

TO SOLVE FOR TSP

1. Copy the right side view where the surface ABCD appears as an inclined line (or as an edge). The inclined line is the edge view (EV) of surface ABCD, as seen from the right.

2. Draw reference line RL parallel to the edge view, and lines s and t perpendicular to RL.

3. Draw AB equal to AB in the front view, and DC equal to DC in the front view.

4. ABCD, in heavy lines, is the desired TSP.

148. 1) Follow steps 1 and 2 in the preceding problem to obtain points F, G, H, and I. FGHI is the desired top view.

Fig. 315

TO SOLVE FOR THE TSP

1. Copy the front view, the view where the inclined surface FGHI is shown. FGHI appears as an edge in the front view. Thus, the edge view (EV) is in the front view.

2. Draw the RL parallel to the EV, then draw lines s and t perpendicular to RL.

3. On line s, draw FI equal to FI in the right side view, and GH equal to GH in the right side view.

4. FGHI is the desired TSP.

149. ABC, above the given front view, is the desired top view. Follow the steps in probs. 146 and 147.

Fig. 316

TO FIND THE TSP.

This problem is similar to prob. 146. Proceed similarly as in prob. 146 to obtain the TSP.

150.   ABC, above the given front view, is the desired top view.
Follow the steps in prob. 147.

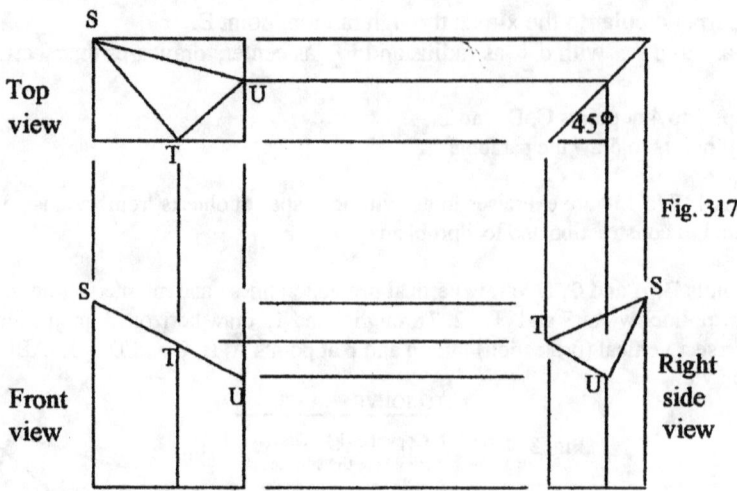

Fig. 317

TO SOLVE FOR THE TSP: Follow the solution to prob. 148.

151.

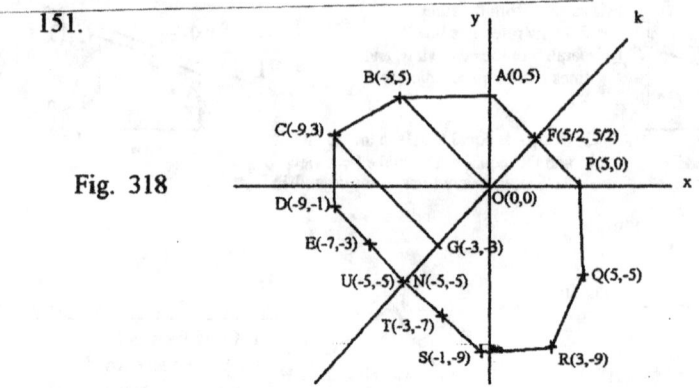

Fig. 318

The distances of A, B, C, D, and E to the line of reflection k are determined by finding the
length of the perpendicular from these points to the point of intersection with k.

The equation of k is y = x and therefore its slope = 1, and the slope of all lines perpendicular to k has to be -1.   This means that the slopes of AF, CG, DU, and EN are equal to 1.

Distance AF:
Eq. of AF: y - 5 = -1 (x - 0).   Eq. of k: y = x   Solving equations simultaneously: x = 5 / 2,
y = 5 / 2.   Coordinates of F: (5/2, 5/2)
By the distance formula: $AF = \sqrt{(0 - 5/2)^2 + (5 - 5/2)^2}$ = 3.54

Distance CG:
Eq. of CG: y - 3 = -1 (x + 9)   Eq. of k: y = x   Solving equations simultaneously: x = -3,
y = -3.
By the distance formula: $CG = \sqrt{(-9 + 3)^2 + (3 + 3)^2}$ = 8.46

Distance DU:
Eq. of DU: y + 1 = -1 (x + 9)   Eq of k: y = x.   Solving equations simultaneously: x = -5,
y = -5.
By the distance formula: $DU = \sqrt{(-9 + 5)^2 + (-1 + 5)^2}$ = 5.66

Distance EU or EN:
Eq. of EN: y + 5 + -1 (x + 5)   Eq. of k: y = x.   Solving equations simultaneously: x = 5,
y = -5.
By the distance formula: $EN = \sqrt{(-7 + 5)^2 + (-3 + 5)^2}$ = 2.83

152.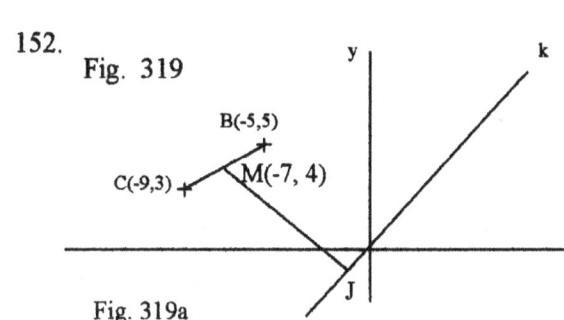

Fig. 319

Fig. 319a

By the midpoint formula, the midpoint of BC is at M(-7, 4); slope of MJ = -1. Thus, the equation of MJ is:
y - 4 = -1 (x + 7)
of y = -x - 3.

Fig. 319b

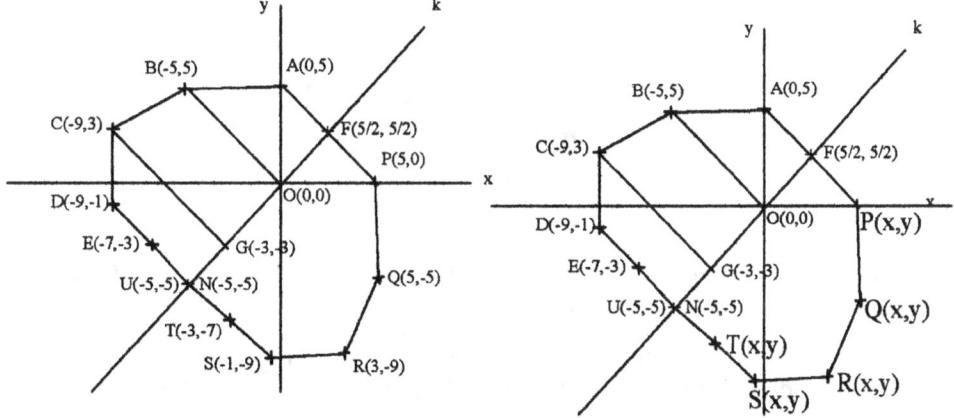

153. Coordinates of P(x, y): (Refer to Fig. 319b).
Since F is the midpoint of AP (by symmetry of the image and the pre-image along the line of reflection k), and by the midpoint formula:
(x + 0) / 2 = 5/2 and x = 5; (y + 5) / 2 = 5/2, y = 0.
Coordinates of Q(x, y):
O is the midpoint of BQ, thus (x - 5)/2 = 0, x = 5; (y + 5) /2 = 0, y = -5.
Coordinates of R:
G is the midpoint of CR, thus (x - 9) / 2 = - 3, x = 3; (y+3) / 2 = -3, y = -9.
Coordinate of S:
U is the midpoint of DS; thus (x - 9) /2 = -5, x = -1; (y - 1)/2 = -9, y = -9.
Coordinates of T:
The midpoint of ET is N; thus (x - 7)/2 = -5, x = -3; (y - 3)/2 = -5, y = -7.

The above coordinates of P, Q, R, S, and T are shown in Fig. 319a.

154. m_{BT} = (5 + 7) / (-5 + 3) = -6. (Refer to Fig. 319a for the coordinates of B and T.

155. The translation maps (x,y) into (x+6, y+10) so that the equation of the translational image is: y + 10 = -2(x+6) + 2 or y= -2x -20 (shown as a dashed or broken line in Fig. 320).

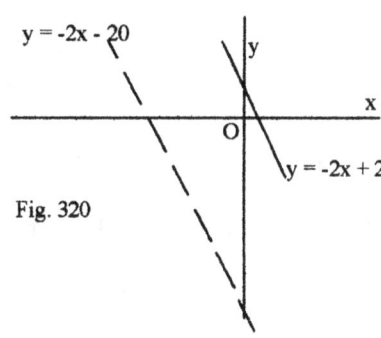

y = -2x - 20

y = -2x + 2

Fig. 320

156.

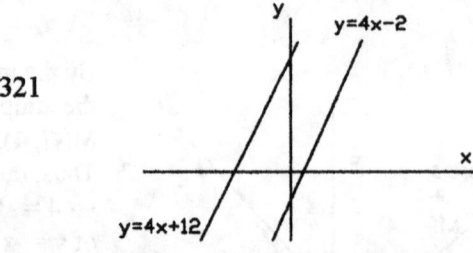

Fig. 321

The point (x, y) is mapped into (x + 6, y + 10) under this translation. The equation of the pre-image is y = 4x-2. The equation of the image: y + 10 = 4(x + 6) - 2 or y = 4x + 12. The image line is shown as a broken or dashed line in Fig. 321.

157.

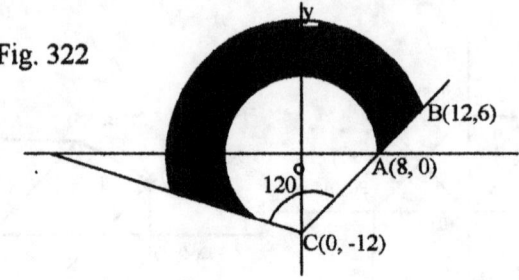

Fig. 322

Area brushed by the wiper blade (the shaded region in Fig. 322) = Area of big sector of a circle - Area of small sector of a circle.

The radius of big circle is: $CB = \sqrt{(12 - 0)^2 + (6 + 12)^2} = 21.63$
The radius of the small circle is $CA = \sqrt{(8 - 0)^2 + (0 + 12)^2} = 14.42$

Area of shaded region
$= (120 / 360) \pi R^2 - (120 / 360) \pi r^2$
$= (1 / 3) \pi (R^2 - r^2) = (1 / 3)\pi [(21.63)^2 - (14.42)^2]$
= 272.91 sq. units.

158.   Fig. 323

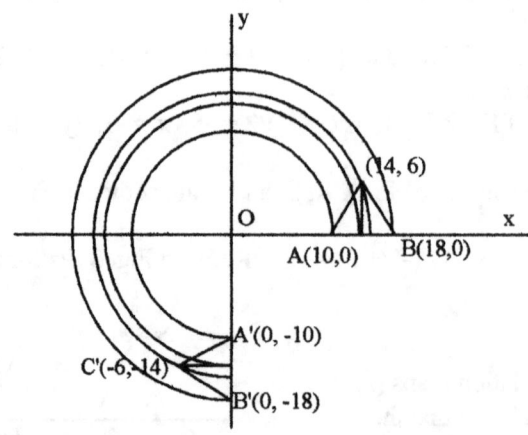

Length of arc AA' = (270/360) 2 π 10 = 47.1 units.

160. Area of sector BB' = (270 / 360)   18   = 7.63 sq. units

161. Area of sector ADA' = (270 / 360)   10   = 235.5 sq. units.

162. (Refer to Fig. 323, prob. 158). Use Hero's Formula to find the area

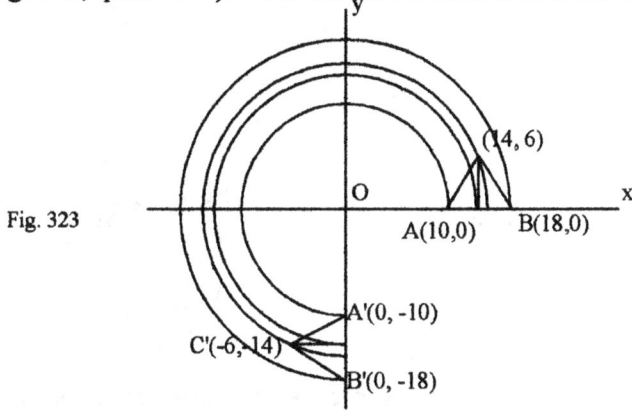

Fig. 323

Find the length of the sides of triangle AA'C':
AA' = $\sqrt{10^2 + 10^2}$ = 14.14
A'C' = $\sqrt{6^2 + 4^2}$ = 7.21
AC' = $\sqrt{16^2 + 14^2}$ = 21.26   s = (14.14 + 7.21 + 21.26) / 2
                                              = 21.31.
Area of triangle AA'C' = $\sqrt{s\ (s-a)\ (s-b)\ (s-c)}$
                       = $\sqrt{21.31\ (7.17)\ (14.1)\ (.05)}$ = 10. 36 sq. units.

163. Refer to Fig. 323.
Length of arc CC' = {[(270 - 2 ∠COB) / 360] 2 $\tilde{\pi}$(10)]}
                    where tan ∠COB = 6 / 14 = .43
                    and where ∠ COB = 23.27 .
Length of arc CC' = 54.57 units.

164. The scale factor k for this dilatioon is k = 3, which means that the original solid expands along all three axes by a factor of 3. Hence the required coordinates for K', L', and M' are: K'(0, 12, 12), L' (12, 12, 12), and M' (12, 12, 0). See Fig. 324

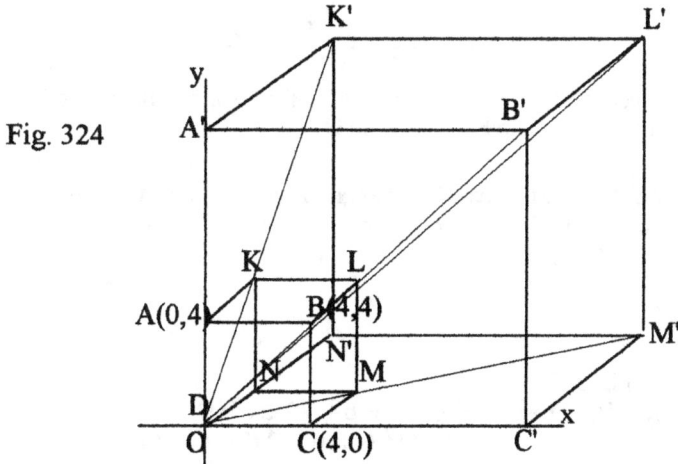

Fig. 324

165. Length of the diagonal of the image cube:
OL' = $\sqrt{12^2 + 12^2 + 12^2}$ = 20.78.

166.

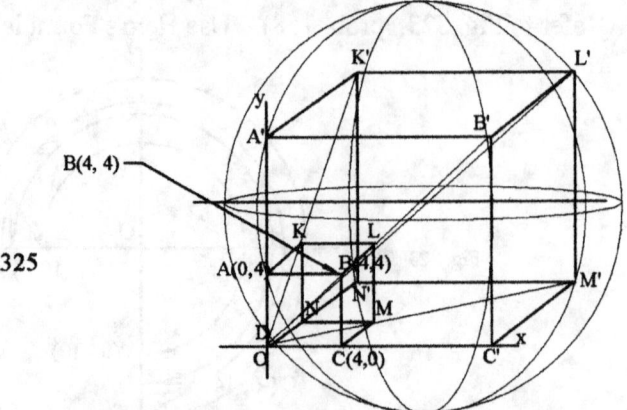

Fig. 325

V_{sphere} = $(4 / 3) \pi r^3$, where r is the radius of the circumscribed sphere
$= (12 \sqrt{3}) / 2$ = 10.38.
= 1.33 (3.14) 10.38^3 = 4670.62 cu. units.

167. Area of inscribed sphere: A = $4 \pi r^2$ = 4 (3.14) 6^2 = 452.16 sq. units.

168. Gale component
in the northerly direction
= 200 cos 42 = 200(.74)
= 148 fps.

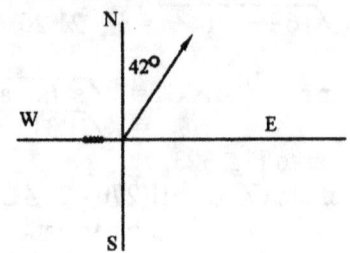

169. By the Law of Cosines,
$c^2 = a^2 + b^2 - 2ab \cos C$
$= 30^2 + (100)^2 - 2 (30) 100 \cos 138$
= 6460
c = 80.37 fps

Fig. 327

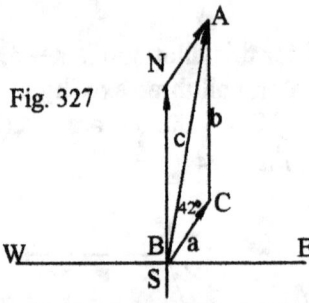

To find the direction of the resultant velocity of the boat, use the Law of Sines: a / sin A = b / sin B = c / Sin C

Our goal is to determine the measure of angle B (which is the angle opposite side b); so   c / sin C = b / sin B;   80.37 / sin 138 = 100 / sin B; sin B = .83, and B = 56°.

170. (Refer to Fig. 328)
Using the Law of Cosines, $c^2 = a^2 + b^2 - 2ab \cos C$
$= 30^2 + 100^2 - 2 (3) 100 \cos 138$
= 6460
c = 80.37

170.

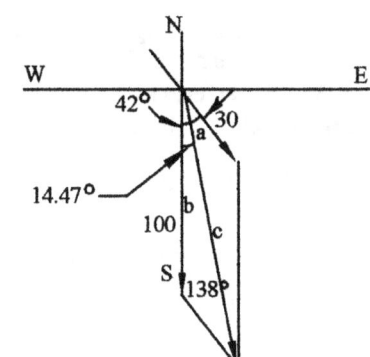

Fig. 328

Use the Law of Sines to find the direction of the resultant velocity:
  a / sin A = c / sin C ;   30 / sin A = 80.37 / sin 138
sin A = .25,   A = 14.47°

171. To find the magnitude of the resultant velocity of the submarine, use again the Law of Cosines.

Fig. 329

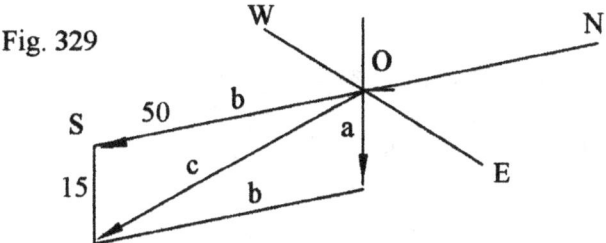

$c^2 = a^2 + b^2 - 2ab \cos C = 15^2 + 50^2 - 2(15)50 \cos 90 = 2725,$
c = 52.2 knots (Note: The Law of Cosines, in this case, reduces to the Pythagorean Theorem because the angle C is 90 degrees, and cos 90 = 0.

Direction of the resultant velocity:
by the Law of Sines,   a / sin A = c / sin C;   15 / sin A = 52.2 / 1
sin A = .29,   A = 16.86°.
Hence, the resultant velocity is directed at 16.86 degrees below the water surface.

172.

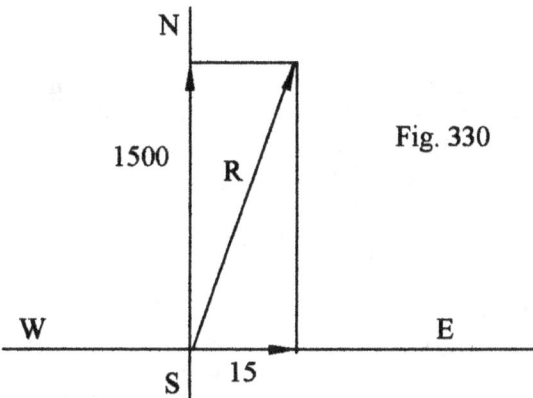

Fig. 330

To find the resultant velocity of the missile: $R^2 = 15^2 + 1500^2$
$R = \sqrt{225 + 2,250,000} = 1500.07$ mph.
Time for the missile to hit the target = 100 / 1500.07 = .07 hr = 252 sec.

173. The unit vectors i, j, and k represent unit components of a given vector in space along the x, y, and z axes respectively. For instance, the diagonal of a cube 6 units on an edge would be represented by $6i + 6j + 6k$. The vector would be as illustrated in the figure below. The magnitude of R (indicated by $|R|$) is:

$$|R| = \sqrt{6^2 + 6^2 + 6^2} = 6\sqrt{3}.$$

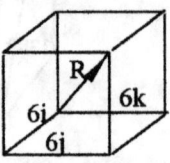

Fig. 331

For the given vectors, $P = 2i + 6j + 4k$, $Q = 4i + 6j + 4k$

$$|P| = \sqrt{2^2 + 6^2 + 4^2} = 7.49 \qquad |Q| = \sqrt{4^2 + 6^2 + 4^2} = 8.25.$$

174. $|R| = \sqrt{16 + 36 + 16} = 8.25$. $|S| = \sqrt{36 + 16 + 16} = 8.25$

175. $|K| = \sqrt{4 + 36 + 16} = 7.48$ $|L| = \sqrt{36 + 36 + 16} = 9.38$

176. $P = 2i + 6j + 4k$ $Q = 4i + 6j + 4k$ $R = 4i + 6j + 4k$

$$|P + Q + R| = \sqrt{10^2 + 18^2 + 12^2} = 23.83.$$

177. From prob. 176. $|P - Q| = \sqrt{(-2)^2 + 0^2 + 0^2} = 2$

178. From prob. 176. $|R - P| = \sqrt{(4-2)^2 + (6-6)^2 + (4-4)^2} = 2$.

179. $R = 4i + 6j + 4k$, $S = 6i + 4j + 4k$

The direction cosines for a space vector are: $\cos\theta_x$, $\cos\theta_y$ and $\cos\theta_z$. They are the components of the unit vector in the direction of a given vector R or S. For example, the direction cosines of R in prob. 176 are:

$\cos\theta_x = 4 / |R|$, $\cos\theta_y = 6 / |R|$, $\cos\theta_z = 4 / |R|$.

Direction cosines of R:

$|R| = \sqrt{4^2 + 6^2 + 4^2} = 8.25$

so, $\cos\theta_x = 4 / 8.25 = .48$

$\cos\theta_y = 6 / 8.25 = .73$

$\cos\theta_z = 4 / 8.25 = .48$

Direction cosines of S:

$|S| = \sqrt{6^2 + 4^2 + 4^2} = 8.25$.

thus, $\cos\theta_x = 6 / 8.25 = .73$

$\cos\theta_y = 4 / 8.25 = .48$

$\cos\theta_z = 4 / 8.25 = .48$

180. $K = 2i + 6j + 4k$

Direction cosines of K

$|K| = \sqrt{2^2 + 26^2 + 4^2} = 7.48$

$\cos\theta_x = 2 / 7.48 = .27$

$\cos\theta_y = 6 / 7.48 = .80$

$\cos\theta_z = 4 / 7.48 = .53$

$L = 6i + 6j + 4k$

Direction cosines of L

$|L| = \sqrt{6^2 + 6^2 + 4^2} = 9.38$

$\cos\theta_x = 6 / 9.38 = .64$

$\cos\theta_y = 6 / 9.38 = .64$

$\cos\theta_z = 4 / 9.38 = .43$

181. $P = 2i + 6j + 4k$ $\qquad$ $Q = 4i + 6j + 4k$

$|P| = \sqrt{4 + 36 + 16} = 7.48$ $\qquad$ $|Q| = \sqrt{16 + 36 + 16} = 8.25$

Direction cosines of P $\qquad$ Direction cosines of Q

$\cos \theta_x = 2 / 7.48 = .27$ $\qquad$ $\cos \theta_x = 4 / 8.25 = .48$

$\cos \theta_y = 6 / 7.48 = .80$ $\qquad$ $\cos \theta_y = 6 / 8.25 = .73$

$\cos \theta_z = 4 / 7.48 = .53$ $\qquad$ $\cos \theta_z = 4 / 8.25 = .48$

182. $P = 2i + 6j + 4k$ $\qquad$ $Q = 4i + 6j + 4k$

By definition, the scalar product of P and Q is: $P \cdot Q = |P| \, |Q| \cos \theta$

$$\cos \theta = (P \cdot Q) / |P| \, |Q| = \frac{(2i + 6j + 4k)(4i + 6j + 4k)}{\sqrt{56} \ \sqrt{68}} = .97$$

$\theta = 14.07^{\circ}$

183. $R = 2i + 6j + 4k$ $\qquad$ $S = 6i + 4j + 4k$

Using the scalar or dot product formula:

$$\cos \theta = (R \cdot S) / |R| \, |S| = \frac{(2i + 6j + 4k)(6i + 4j + 4k)}{\sqrt{56} \ \sqrt{68}} = .84$$

$\theta = 32.86^{\circ}$

184. $K = 2i + 6j + 4k$ $\qquad$ $L = 6i + 6j + 4k$

$$\cos \theta = (K \cdot L) / |K| \, |L| = \frac{(2i + 6j + 4k)(6i + 6j + 4k)}{\sqrt{56} \ \sqrt{88}} = .91$$

$\theta = 9.54^{\circ}$

185. To find the dot (or scalar) product of A (x_1, y_1, z_1)
and B (x_2, y_2, z_2): $A \cdot B = (x_1 + y_1 + z_1)(x_2 + y_2 + z_2)$
For $K = 2i + 4j + 0k$ and $L = 6i + 4j + 4k$
$K \cdot L = 12i^2 + 8ij + 8ik + 24ij + 16j^2 + 16jk + 0ik$
$\qquad$ $+ 0kj + 0k^2$)
$\qquad$ $= 12i^2 + 16j^2 + 0k^2 = 12i + 16j = 28.$
(Note: $i^2 = 1$, $j^2 = 1$, and $k^2 = 1$, the reason being that these
vectors have a zero angles between them and that $\cos 0 = 1$; each
of the ij, ik, and jk terms become 0 since they are at right angles
to each other and that $\cos 90 = 0$).

186. $P = 2i - 3j + k$; $\qquad$ $Q = -i + 2j - k$
As in the preceding problem, the dot product of P and Q is:
$P \cdot Q = -2i^2 - 6j^2 - k^2 = -6 -2 -1 = -9.$
(Note that the product of P and Q should have 9 terms, but only three remain
and these are the i^2, j^2, and k^2 terms. The ij, ik, and jk terms reduce
to 0, since they are at 90 degrees to each other and that $\cos 90 = 0$.

187. The cross product, or vector product, of two vectors A and B is the vector perpendicular to the plane through A and B, and is defined as:
$A \times B = |A| \, |B| \sin \theta$ (where θ is the angle between A and B).
Note: The cross product is positive if it is in the same direction as the thumb
pointing up and the other fingers grasping the vector, negative when the thumb
is pointing down).

See Fig. 332.

$K = 4i - 6j - 6k$ $L = 8i + 4j + 2k$

$$K \times L = i \begin{vmatrix} i & j & k \\ 4 & -6 & -8 \\ 8 & 4 & 2 \end{vmatrix} - j \begin{vmatrix} i & j & k \\ 4 & -6 & -8 \\ 8 & 4 & 2 \end{vmatrix} + k \begin{vmatrix} i & j & k \\ 4 & -6 & -8 \\ 8 & 4 & 2 \end{vmatrix}$$

$$= i \begin{vmatrix} -6 & -8 \\ 4 & 2 \end{vmatrix} - j \begin{vmatrix} 4 & -8 \\ 8 & 2 \end{vmatrix} + k \begin{vmatrix} 4 & -6 \\ 8 & 4 \end{vmatrix}$$

The rows and columns CROSSED OUT by the broken lines in the upper determinant leave only the terms in the lower determinant.

$K \times L = i(-12 + 32) - j(8 + 64) + k(16 + 48) = 20i - 72j + 64k$

The direction of the cross product is determined by the right-hand rule:

The signs for i, j, and k are alternating plus and minus (+, -, and +) respectively.

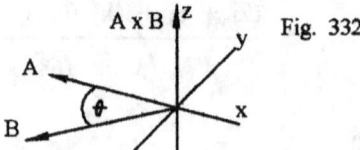
Fig. 332

If the fingers of the right hand curl in the direction from A to B (see Fig. 332), the thumb points in the direction of the cross product.

188. Find the cross product $L \times K$. $K = 4i - 6j - 8k$ $L = 8i + 4j + 2k$

$$L \times K = i \begin{vmatrix} i & j & k \\ 8 & 4 & 2 \\ 4 & -6 & -8 \end{vmatrix} - j \begin{vmatrix} i & j & k \\ 8 & 4 & 2 \\ 4 & -6 & -8 \end{vmatrix} + k \begin{vmatrix} i & j & k \\ 8 & 4 & 2 \\ 4 & -6 & -8 \end{vmatrix}$$

$$= i \begin{vmatrix} 4 & 2 \\ -6 & -8 \end{vmatrix} - j \begin{vmatrix} 8 & 2 \\ 4 & -8 \end{vmatrix} + k \begin{vmatrix} 8 & 4 \\ 4 & -6 \end{vmatrix}$$

$$= -20i + 56j - 64k$$

189. $P = 2i - 3j + k$
$Q = -i + 2j - k$
$R = 0i + 0j + 4k$

$$Q \times R = i \begin{vmatrix} 2 & -1 \\ 0 & 4 \end{vmatrix} - j \begin{vmatrix} -1 & -1 \\ 0 & 4 \end{vmatrix} + k \begin{vmatrix} -1 & 2 \\ 0 & 0 \end{vmatrix}$$
$$= 8i + 4j + 0k$$

$P \cdot (Q \times R) = (2i - 3j + k)(8i + 4j + 0k)$
$$= 16i^2 + 8ij + 0ik - 24ij - 12j^2 - 0jk + 8ik + 4jk + 0k^2$$
$$= 16 - 12 = 4$$

(Note: All ij, ik, ji, jk, ki and kj terms reduce to 0, since cos $90 = 0$; also, $i^2 = j^2 = k^2 = 1$, since cos $0 = 1$.

190. P · Q = (2i - 3j + k) (-i + 3j - k)
$$= 2i^2 - 4ij - 2ik + 3ij - 6j^2 + 6jk - ik + jk - k^2$$
$$= -2 - 6 - 1 = -9$$

(P · Q) x R = (-2i - 6j - k) (0i + 0j + 4k)

$$\begin{vmatrix} i & j & k \\ -2 & -6 & -1 \\ 0 & 0 & 4 \end{vmatrix} = i \begin{vmatrix} -6 & -1 \\ 0 & 4 \end{vmatrix} - j \begin{vmatrix} -2 & -1 \\ 0 & 4 \end{vmatrix} + k \begin{vmatrix} -2 & -6 \\ 0 & 0 \end{vmatrix}$$

$$= 24i + 8j$$

191. (Fig. 333) For the system to be in equilibriuim, the sum of the moments in the clockwise direction must be equal to the sum of the counterclockwise moments. That is, the algebraic sum of the clockwise and counterclockwise moments must be equal to 0. The technical notation for this statement is: $\Sigma_{M_B} = 0$.

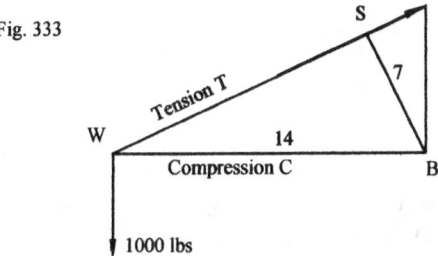

Fig. 333

Clockwise moment (s) :   1000 x 14
Counterclockwise moment: T x 7   where 7 is the perpendicular distance of the line of action of T from B.
1000 x 14 = T x 7 ; T = 14000 / 7 = 2000 lbs.

192. To find the compression force on C on the horizontal beam in the system (Fig. 333). Two other conditions for the system to be in equilibrium would be that: 1) The forces acting to the right must be equal to those acting to the left, or that the algebraic sum of horizontal forces must be equal to 0; and 2) The forces acting upward must equal the forces acting downward. That is the algebraic sum of vertical forces must equal 0.

Using condition (1) to solve for the compression on the beam:
The horizontal component of T, which is T cos 30, is directed to the right, adn the compression at C pushes the cord to the left, hence:
     Compression C = T cos 30,   thus C = 2000 (,87) = 1740 lbs.

193. For the system in Fig. 334 to be in equilibrium, the algebraic sum of horizontal forces must be equal to 0, and the algebraic sum of vertical forces must be equal to 0. Simply stated, the upward forces must equal the downward forces   and the forces to the right must be equal to those directed to the left.

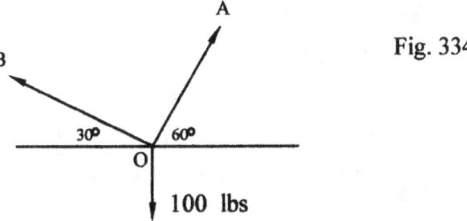

Fig. 334

Forces to the right must equal forces to the left:
OA cos 60 = OB cos 30; .5 OA = .87   OA = 1.74 OB.

The sum of vertical forces upward must equal vertical forces going down:

OA sin 60 + OB sin 30 = W

OA (.87) + OB (.5) = 100 lbs

(1.74 OB (.87) + .50 OB = 100 ;   1.51 OB + .50 OB = 100;

OB = 49.75 lbs, OA = 1.74 OB = 86.57 lbs.

194. For the system to be in equilibrium, the clockwise moments must be equal to the counterclockwise moments.

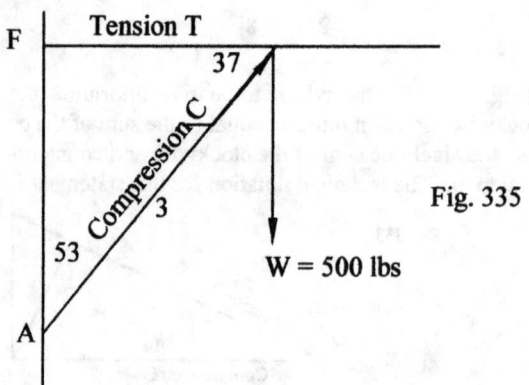

Fig. 335

Taking moments about pt A: Clockwise moments must equal counterclockwise moments.   W (3 sin 53) = T (3 sin 37)

500 (3) .8   = T (3) .6   T = 1200 / 1.8 = 666.67 lbs

To find the compression force C: Forces acting up must equal forces acting down.

C cos 53 = W ;   C (.6) = 500 ;   C = 500 / .6 = 833.33 lbs.

The polar form of the complex number C is: | C | $\underline{/\theta}$ where | C | is the

195. Complex numbers have two components, namely: a real number and an imaginary number, and is written in the form A +Bi or A + iB, where A is the real part and iB the imaginary part.

Complex numbers can be expressed in either of two ways: Cartesian coordinates or with polar coordinates.

In the given complex number -9 -i6, the real part is -9 and the imaginary part is i6. That is, A = -9, and B = -6.   tan θ = B / A = -6 / -9 = .67. θ = 33.82°.

magnitude of the vector from the origin (0, 0) to the point C. The angle θ is the counterclockwise angle formed by vector C and the real axis.

| C | $= \sqrt{(-9)^2 + (-6)^2} = \sqrt{117} = 10.82$.

Thus, the polar form of the complex number -9 -i6 is C = 10.82 $\underline{/33.82}$ °

196. Given, 50 + i40.

| C | $= \sqrt{(50)^2 + (40)^2} = 64.29$.   tan θ = 40 / 50 = .8 ; θ = 38.66°.

In polar form, C = 64.29 $\underline{/38.66°}$

197. Given, 10 - i4 ; | C | $= \sqrt{(10)^2 + (4)^2} = 10.77$; tan θ = -4 / 10 = - .4 ; θ = 21.8° In polar form, C = 10.77 $\underline{/21.8}$ °

198. Given, 7 $\underline{/120°}$; tan 120 = -1.73 = - $\sqrt{3}$ / 1 . Hence, in Cartesian form, C = 1 - i$\sqrt{3}$.

199. To find the roots of x^2 + 6x + 12 = 0. By the quadratic formula,

$x = \dfrac{-6 \pm \sqrt{6^2 - 4\,(1)\,12}}{2\,(1)} = -3 \pm i\,\sqrt{3}$. The roots are - 3 + i $\sqrt{3}$

and - 3 - i $\sqrt{3}$.

In polar form:

$A = -3, B = \sqrt{3}$. $|C| = \sqrt{A^2 + B^2} = \sqrt{12} = 3.46.$

$\tan \theta = 3 / -3 = -1/3$; $\theta = 30°$, $= -30°$ Hence, $C = 3.46 \underline{/\pm 30°}$

200. $(3 + i6) \times (10 - i7) = 30 + i60 - i21 - i^2 42)$ (Note that $i^2 = -1$)

= $72 + i39$.

201. $(-6 - i5) \times (8 + i8) = -48 - i40 - i48 - i^2 40 = -48 - i88 + 40$

= $-i - i88$.

202. $(10 - i2) \times (-16 + 17) = -160 + i32 + i70 - i^2 14 = -146 + i102$

203. $\dfrac{(3 + i6)}{(10 - i7)} \cdot \dfrac{(10 - i7)*}{(10 - i7)*}$ (The asterisk * indicates that the expression is the conjugate of the denominator of the fraction at the left).

$= \dfrac{30 + i60 - i21 - i^2 42}{100 - i^2 49}$

$= -12/149 + i72/149$

204. $\dfrac{(-6 - i5)}{(8 + i8)} \cdot \dfrac{(8 + i8)*}{(8 + i8)*} = \dfrac{-48 - i40 - i48 - i^2 40}{64 - i^2 64} = \dfrac{-8 - i88}{64 + 64}$

= $-8/128 - i88/128$ = $-1/16 - i11/16$

205. $\dfrac{(10 - i2)}{(-16 + i7)} \cdot \dfrac{(-16 + i7)*}{(-16 + i7)*} = \dfrac{-160 + i32 + i70 - i^2 14}{256 - i^2 44}$

$= \dfrac{-146 + i102}{300} = -146/305 + i102/305$

206. $7 - i4$

207. $-12 - i15$

208. $8 + i8$

Problems Index

www.ingramcontent.com/pod-product-compliance
Lightning Source LLC
Chambersburg PA
CBHW081140170526
45165CB00008B/2739